The
Telecom
Tutorials

by Jane Laino

**A Practical Guide
for Managing Business
Telecommunications Resources**
2nd Edition

Published by CMP Books
An Imprint of CMP Media Inc.
12 West 21st Street
New York, NY 10010

ISBN 1-57820-093-8

For individual orders, and for information on special discounts for
quantity orders, please contact:
CMP Books
6600 Silacci Way
Gilroy, CA 95020

Tel: 1-800-500-6875 or 408-848-3854
Fax: 408-848-5784
Web: www.cmpbooks.com
Email: cmp@rushorder.com

Distributed to the book trade in the U.S. and Canada by
Publishers Group West
1700 Fourth St., Berkeley, CA 94710

Manufactured in the United States of America

The Telecom Tutorials

Jane Laino, author of the widely acclaimed Telecom Handbook, presents the second expanded edition of The Telecom Tutorials. This updated series of brief "how to" guides enables the neophyte to get quickly up to speed in navigating the pitfalls of managing business telephone systems and services. Seasoned telecommunications professionals will also learn new tips and benefit from the practical perspectives that can only be obtained by day-in, day-out exposure to real world businesses telecommunications issues.

Tutorials include guides for...

▶ Understanding the United States Telecommunications Industry.

▶ Purchasing Local and Long Distance Telephone Service and Business Telephone Systems.

▶ Setting Up Your Telephone Systems to Provide Optimal Service to Your Staff and to Callers.

▶ Managing Telecommunications Resources to keep systems running smoothly and records up-to-date.

▶ Staffing for Telecommunications Support.

▶ Reducing and controlling Telecommunications Expenses.

▶ Understanding the "nuts and bolts" of telecommunications systems.

▶ Understanding some advanced technologies for the delivery of telecommunications services.

You can read this book straight through or jump right to the sections you need most. Each tutorial stands alone as a helpful, non-technical guide to its topic.

What Readers Are Saying
About The Telecom Tutorials

These letters were received as a tutorial appeared in the monthly publication called TELECONNECT which has now been merged with Communications Convergence Magazine (from CMP Media).

> "I really appreciate your tutorial on DSL. It is by far the best article I have seen on the topic."
>> *Alan H. Baldwin, Commtran*

> "I read your guide 'Tour of the Phone Room' and loved it! It helped me to understand more about the room since I inherited the phone responsibilities without any training."
>> *Stephanie Martin, Maytag*

> "I really like the way you explain telecom in plain terms."
>> *David M. Wylie, Bank of New York*

> "With great interest I read your article "Is your auto attendant easy for your callers to use?" as we are makers of a PBX system with auto attendants. Our engineers were happy to see that we met most of your criteria!"
>> *Marc Rand, Centrepoint Technologies*

> "Wow! Your article about traffic reports was very informative & timely."
>> *John Saavedra, U.S. Postal Service*

> "I read so much clap trap it was a pleasant surprise to see your article which makes a complex subject simple and understandable."
>> *Jim VanHorn, Educator*

Do You Fall Into One Or More Of The Following Categories?

If so, this book will help you.

▶ **Business Owners and Managers** who need to understand what's behind their most critical business systems.

▶ **Telecommunications Companies** of all types who need to train staff members and keep employees current.

▶ **Communications Convergence Professionals** who need to know what organizations expect from their telephone systems.

▶ **Information Technology Professionals** who have inherited the responsibility for the telephone system and are learning that it is very different from the computer systems.

▶ **Students and Educators** who will supply the workforce with needed telecommunications expertise in the 21st century.

▶ **Investors in Telecommunications Companies** who must understand the terminology of this changing industry in order to make good decisions.

▶ **Bookstore Owners and Managers** whose shops offer rows of computer books for sale — but no books on telephone systems and services.

▶ **Professionals in Related Disciplines**, such as engineering, who need a working knowledge of voice communications.

Read the Table of Contents for more insight.

For tips on managing telephone systems, services and expenses visit DIgby 4 Group, Inc.'s web site at www.digby4.com

**This book is dedicated
to my Aunt and Uncle**
Helen and Jim Canvisser

Thanks For Your Help Everyone

Christine Kern of CMP Books who supported the idea behind the original and new edition of this book, compiling a monthly TELECONNECT Magazine feature into book form. Thanks for being a pleasure to work with, Christine!

Warren Hersch, Elaine Rowland (now Elaine Kahan) and **John Jainschigg,** all past editors of TELECONNECT Magazine for providing insight, guidance and an outstanding forum for the tutorials. Warren is now editor of **Call Center Magazine** and John is editor of **Communications Convergence**, both CMP Publications.

Harry Newton founder of TELECONNECT Magazine who inspired each of us who wrote for the magazine along with countless telecommunications professionals. We miss you, Harry!

Joc Petrella of Paradyne; **Adrianne Davis, Stephanie Nicoll, Bob Dorskind** and **Peter Parinello** of InfoHighway; **Cari C'deBaca** of Cisco Systems, Inc.; **Walter Karopczyc** and **Rich Cabelo** of Bristol Capital, Inc. and **Jerry Harder**, traffic engineering specialist for their help with some of the tutorials.

The staff of DIgby 4 Group, Inc. including **Diane Ventimiglia, Jean Fitzpatrick, Laura Taylor** and **Lisa Schalk** for keeping our company running smoothly and our clients happy.

Pattie Stone for her patience and flexibility, not to mention her skills in the design and layout of the book's contents.

My grammar school and high school English teachers in the Bayville-Locust Valley, NY school system who taught me how to write! **Ellen Dunn, Joan Walker, Antionette Savino, Anna Grace Oslansky, Caroline L. Wilkins, Camilla Sperandei, Miss Rennebaum, Paula Webb, Roberta Lee Bishop, George Ball** and **Kathryn Copeland.**

Special thanks to my husband **Rich Laino** who supports all of my efforts and wishes they took up less of my time!

About the Author

Jane Laino is president of DIgby 4 Group, Inc., a telecommunications consulting company providing services to a growing base of business clients.

DIgby 4 helps with the purchase and management of telecommunications systems and services. DIgby 4 clients are comfortable that their systems and services are being purchased at the right price and are appropriate for how they plan to use them.

Jane's history in the telecommunications industry includes seven years with the Bell System, New York Telephone and Southwestern Bell (now called Verizon and SBC) and four years with NatCom, Inc. In 1979 she founded DIgby 4 Group, Inc.

She is known for having a practical, hands-on approach to consulting which translates to the style from which the clients of DIgby 4 Group, Inc. benefit.

She is a member of the Society of Telecommunications Consultants and COMP-TIA. Her speaking, teaching, writing and consulting engagements keep DIgby 4 Group up to date and actively learning in the fast moving telecommunications industry.

Jane lives in New York City during the week and in Southampton, NY on the weekends. She is married to Richard Laino. She was born in Jersey City, NJ and grew up in Bayville, Long Island, NY. She is a graduate of Queens College of the City of New York with a B.A. degree.

In addition to this book, she is author of The Telecom Handbook, now in its 4th edition.

Table of Contents

SECTION 2
Purchasing Telecommunications Equipment and Services
You don't really enjoy tearing up dollar bills —do you? 41

▶ **Tutorial #3**
Purchasing Local Telephone Service –
In the beginning there was the dialtone... 43

▶ **Tutorial #4**
Ordering and Provisioning Telecommunications Services –
To be patient is a virtue. To be detail-oriented is divine!... 51

Tutorial #5

Negotiating Next Year's Long Distance Contract –
Not for people with previous history of heart problems! ... 63

Tutorial #6

Purchasing Calling Card Services – *Beware of the Bong!* .. 69

▶ Tutorial #16
Setting Up A Small Incoming Call Center –
Find Out How Long Your Callers Are Willing To Wait! ... 135

▶ Tutorial #17
"Right-Sizing" Your Telephone System –
Not too big, not to small 143

Tutorial #18
Provisioning the Surround Stuff –
Some callers may prefer automated systems to

SECTION 4
The Finer Points of Managing Telephone Systems –

Tutorial #19
How to Manage Your Telephone System –

SECTION 8
A Little Advanced Technology –
"Optional – Only if you really want to know." **271**

The United States Telecommunications Industry

Read fast —
Before it changes again!

Historical Perspective on The United States Telecommunications Industry

The Saga Continues...

In order to understand today's telecommunications industry and services, it helps to have a perspective on how things have developed.

On March 10, 1876 Alexander Graham Bell invented the telephone. He filed his application for a patent just hours before his competitor, Elisha Gray.

While we're not going to cover it all here, the history of telephony is a fascinating story. In the beginning, with multiple local telephone companies competing, you may have had three separate telephone instruments on your desk to call people who subscribed to one of the three different services.

AT&T (American Telephone & Telegraph) was formed in 1885 to provide long distance service to the local telephone service provider known as Bell. In 1900 AT&T became Bell's parent company through a stock purchase. AT&T came to be known as Ma Bell and AT&T owned telephone companies were known as the Bell System.

The Telecommunications Act of 1934 was adopted in order to regulate the AT&T monopoly over telephone service. It created the FCC (Federal Communications Commission) and granted them the authority to review mergers and acquisitions between telephone companies and the power to regulate interstate telephone services.

Intrastate telephone service was left up to the state utility commissions which ultimately led to the creation of monopolies for local telephone service. In some areas of the U.S., the local monopoly was not part of the Bell System, but an Independent Telephone Company such as was GTE or one of many smaller

"Mom and Pop" telephone companies that still exist, particularly in rural areas.

AT&T had already cornered the long distance and local market when the 1934 act took effect. During the 1930's it seemed to make sense that these companies remain monopolies for the protection of consumers, keeping prices in check.

Then the telecommunications industry started to develop and claims of unfair competition against AT&T began. Dick Kuehn, a well-known telecommunications consultant, describes this time in his column in Business Communications Review Magazine (www.bcr.com):

"Back in 1956, an obscure company named Hush-A-Phone began manufacturing a plastic cup-like device that clipped onto a telephone mouthpiece to block extraneous noise. Hush-A-Phone eventually secured a Supreme Court ruling allowing them to attach that device to phones provided by the local telephone company. This paved the way for today's competitive telephone equipment industry, but Hush-A-Phone remained obscure, even after the court ruling.

Then a gentleman named Tom Carter, from Gun Barrel, Texas, used the Hush-A-Phone case as a basis for an anti-trust case against the Bell System. His Carterfone device used a technique called acoustic coupling that enabled a mobile radio to talk to a telephone or vice versa. Carter was looking for a way to help people working in the Texas oil fields, but AT&T and Southwestern Bell (the local telephone company in Texas) had a tariff that restricted the attachment of any device to telephone company property. Carter's anti-trust suit eventually found its way to the U.S. Supreme Court and then to the FCC for resolution in 1969. This was the first time that telephones and business telephone systems could be purchased and owned, rather than rented from the telephone company.

Around the same time, Jack Goeken, a GE mobile radio distributor in Joliet, Illinois, wanted to construct a microwave network between Chicago and St. Louis that his customers could use to dispatch cars and trucks. AT&T, Illinois Bell (local telephone company in Chicago), Southwestern Bell (local telephone company in St. Louis) and Western Union (company who sent telegrams) all opposed that license, so Goeken amended his application to become a specialized common carrier. He'd provide services that weren't available from the traditional carriers. The filing by Goeken's company – Microwave Communications Inc. or MCI– produced the FCC's 1969 decision creating the competitive long-distance industry.

Who would have thought that opposition to a microwave license would set in motion a chain of events that would lead to the breakup of the Bell System. Some years after the FCC's decisions, an AT&T attorney who participated in both cases was asked why neither decision was appealed. His answer was that no one at AT&T ever thought they would lose the case, so when the Order came down against them, they did not have enough time to marshal forces for an appeal.

There have been plenty of other 'who would have thought' situations in the telecommunications industry. During the mid-1970's Northern Telecom (now called Nortel) and ROLM (now Siemens) introduced the first digital telephone systems and the battle was on for what would be the 'office controller.' The other contenders were IBM with its mainframe model and a newcomer, a technology called Ethernet, that was being promoted by DEC (Digital Equipment Corp.) and Xerox. Who would have thought that the newcomer could defeat such entrenched opponents?

And who would have thought that AT&T would shed its equipment arm, Western Electric, which became Lucent and has split again with the entity manufacturing business telephone systems now being Avaya. Who knew that the Regional Bell Operating Companies, created by the AT&T divestiture, would outperform AT&T and wind up as the dominant carriers? And who would have thought, as the debate raged over the Telecommunications Act of 1996 (described below) that competition for local telephone service would be so unsuccessful?"

In 1982, after the Department of Justice filed an antitrust suit against AT&T, what is known as the Modification of Final

Judgement (MFJ) was reached. Under the MFJ, AT&T was to divest its twenty-two existing Bell Companies into seven independent Regional Bell Operating Companies (RBOCs – pronounced "R-box"). AT&T agreed to the break-up of the Bell companies on the condition that it could provide long distance service with almost no restrictions. The RBOCs were completely restricted from offering long distance service, information services and the manufacturing of telecommunications equipment.

In January of 1984, the culmination of this antitrust suit over which Judge Harold Greene presided resulted in the divestiture of the Bell System companies by AT&T. This marked the beginning of more change in the telecommunications industry. New players came in to sell systems and services.

The RBOCs, also known as Baby Bells, were the result of this break up of AT&T. Although AT&T retained its long distance network and the capability to sell business telephone systems, it gave up the ownership of the regulated local telephone companies which became part of the newly formed RBOCs. There are still separate telephone companies today regulated by the state utility commission. Each provides local telephone service (primarily dial-tone lines), handles local telephone calls and switches long distance telephone calls to the appropriate carrier. Separate companies within the RBOC may sell business telephone systems and other telephone equipment, although not all do. Since their inception, the RBOCs have had to surmount regulatory restrictions to obtain the right to compete in existing markets, including long distance telephone calling and other emerging markets, such as the sale of information.

Most services sold by the local telephone companies continue to be tariffed. Although a movement toward detariffing is underway. Tariffing entails a review by a state utility commission. The tariffs are incorporated into a voluminous set of written service descriptions and prices. Any time there is a price increase or a new service is introduced it is subject to review by the commission.

Although the 1984 divestiture ended AT&T's monopoly in the telecommunications industry, competition had not materialized as had been expected. As technology continued to develop in the 1990's, the MFJ's process for allowing RBOCs to offer new telecommunications services was becoming obsolete. Procedures under the rules created numerous legal delays. For example, the RBOC had to go through a process to get permission to offer any new services to its customers. This was a long, drawn out procedure in which the RBOC had to file its reasons with the District Court as to why it should be able to expand services to consumers without hindering competition. This process under the MFJ caused the United States to fall behind other countries in technological innovation. In addition, the MFJ decreased long distance competition, since it banned the RBOCs from providing it.

The shortcomings of the MFJ resulted in Congress updating the 1934 act by passing the Telecommunications Act of 1996. This act attempted to level the playing field for competition in the local telephone service market and enabled RBOCs to offer long distance service. Part of the Act applies to all telecommunications service providers, including the RBOCs. It states that it is the "general duty of telecommunications carriers is to interconnect directly or indirectly with other telecommunications carriers." It also states that they must (1) resell services on reasonable terms (2) provide telephone number portability (3) permit competitors to have access to telephone numbers, operator assistance and directory assistance without unreasonable delay (4) give competitors access to telecommunications equipment such as poles and conduits and (5) they must provide reciprocal compensation for the transport and termination of telecommunications. One of the results of this is that the new competitors to the traditional local telephone company can rent cables from them to deliver the competing services.

Another part of the Telecommunications Act of 1996 permits an RBOC to offer long distance services within its home region providing it complies with safeguards outlined in the Act (in terms of allowing local competition). There are also provisions enabling the RBOCs to sell information (content) and to manufacture telecommunications equipment.

In the years since the 1996 Act was signed into law, there has been a dramatic consolidation of telecommunications companies and these mergers are continuing to take place among RBOCs and the long distance companies. Rates for telecommunications services have dropped considerably.

The face of the telecommunications marketplace changes regularly and usually not in the way that was anticipated by the regulators, the providers of equipment and services and the customers. It's often confusing and contradictory, but never dull!

Telecommunications Industry History: A Personal Perspective From The Author

Many founders of today's telecommunications industry came out of the Bell System and the development of the industry was shaped by their experience. The Bell System was essentially the telecommunications industry in the U.S. for many years. Here's what it was like working for the Bell System from the late 1960's through the mid 1970's.

Some people worked in the craft or plant department end of the business and were members of the Communications Workers of America union (CWA). Installers, repairmen, field foremen, inside foremen, linemen; these were the workers on the front lines that were making it all happen. If you opened up a new business and ordered a telephone line and one telephone, the installer would show up on the due date, hook up the line, test it and install the telephone. It seemed deceptively simple. The installer usually had the order in his hand, written mostly in cryptic codes called USOCs still in use today.

A copy of that same order document was actually received and acted upon by forty different people within the telephone company. When you order a telephone line or even make a simple change in your telephone service today, the process is much the same. Many different departments have roles in the deployment of even the most basic services.

The commercial department included the what was called the business office. The people in this department were members of the Union of Telephone Workers (UTW). I started out at Forest Hills Commercial in Queens, a borough of New York City as a Service Representative. The business office was a structured environment, like the rest of the Bell System. Very little was left to chance.

Before starting to work, I was sent to a month long training class with a group of other women (no men worked in the business office then, and no women worked out in 'the field' doing repairs or installations). The Bell System was big on training. The companies that used to be a part of it still are. We were trained exactly how to react and precisely what to say and do under every conceivable circumstance. If an inconceivable circumstance came up, we were to immediately put the caller on hold and head up to the desk of our "BOS" (Business Office Supervisor), the "Mother" of our unit. Each unit within the business office had approximately six representatives managed by a BOS. For every six units there were six BOS's and a District Manager to manage them. The District Managers reported to a hierarchy of other managers. There was a similar hierarchy in the Plant Department. The foremen were first line, second line, third line, etc., each line having a corresponding set of responsibilities, clout and compensation.

Back in the business office, our job was to take customer orders over the telephone. Orders were written in pencil on paper forms (no computers on the desktop). We released them to the order reviewers, who reviewed them for accuracy. They sent them to the order writers, who typed them into the system for distribution to the other departments. It was also our responsibility to answer customers' questions about their telephone bills and, when we had time, to make collection calls to customers who were late in paying. These were known as "treatment calls."

Orders had to be written on the form in a very precise way, block printing and using the correct USOCs (Universal Service Order Codes) which were consistent throughout most of the Bell

System. The terms CV for a single-line telephone and KV for a six-button telephone rarely heard today were USOC codes.

Another group supporting the business office was the service observers. The observers were hidden away in a nearby unmarked office. Throughout the day, they listened in on a random sampling of calls for each unit. This was to ensure that a high quality of service was maintained and that accurate information was being given out. The worst thing that could happen to a unit was to get a scoring, meaning that the service observer had heard something wrong. The observer would come racing into the business office and post the scoring on a bulletin board for all to see. Naturally, the members of the unit, including the BOS, were humiliated for the remainder of the day. We never knew which of the six members of the unit got the scoring.

There were different types of scorings. One was CWI (Customer Waiting Interval), meaning that a representative had left a customer on hold for an unacceptable amount of time. Another was ICR (Incorrect Rate), usually a result of someone adding the component rates of an order incorrectly (by hand – there were no adding machines on our desks). Another was IHM. I can't remember exactly what it stood for (Inhospitable Manner?), but it meant that a representative had handled the customer in a less than pleasant way. This was the worst type of scoring to get and one that promoted the most speculation about who did it and what was said.

When I transferred from New York Telephone to Southwestern Bell, first in St. Louis and then in Kansas City, the business offices were carbon copies of the one I left behind in New York. Nevertheless, each required a month in a training class before I was put back on the desk to talk to the customers.

The old Bell System was just that, a system and a very good one. The rigidity of structure and customer service monitoring worked well for a company which was regulated by the government and, as a monopoly, had no competition. As competition emerged, first for telephone systems, then for long distance service and finally, for local service, the Bell System's strong but inflexible structure became its weakness.

The layers of management made it difficult to reach a decision quickly. The uniformity of products and procedures throughout the United States prevented different parts of the organization from responding to conditions in the local marketplace. The tariffing of all services required that the Bell System obtain approval from each state's public utilities commission prior to offering a new service. Brand new Custom Calling Services available from home telephones were introduced in 1969: Call Waiting, Call Forwarding, Speed Calling and Three Way Calling. More than thirty years later, the local telephone companies are still offering these same four services as "advanced" capabilities. And the business office is still pretty much the same as it was in the late sixties. The main difference is that customer records are now accessed by computer instead of paper files.

As monopolies, the Bell System telephone companies were regulated cost-based businesses. If they wanted to raise rates, they had to obtain approval from the government regulatory bodies and first had to make a case for why the rate increase was needed. The notion of making large profits did not exist, since it was not permitted. Thus, the Bell System was not the best training ground for the development of entrepreneurial skills. Many years after the AT&T divestiture of the Bell System companies and the deregulation of the telecommunications industry, some employees of the former Bell System companies still think of their business as it was. They have been challenged to adjust it to keep up with the changing marketplace and increased competition.

Getting back to the telephone company business office in the 1970's, if customers hinted that they were going to buy their own telephone system, we had a special "hot line" number to call to report this immediately. Someone from the telephone company sales department would then call to talk the customer out of making this big mistake connecting "foreign" equipment to the Bell System lines. I can remember getting queasy at the thought of someone being foolish enough to do this. That good old Bell System training included a fair amount of brainwashing as well!

The world of telecommunications began to change. Not everyone's entrepreneurial skills had been dulled by the Bell System. More business people were discovering the economic advantages of purchasing telephone systems rather than renting from Ma Bell.

In 1976, the interconnect industry was taking off thanks to the way being paved by Tom Carter several years earlier. Interconnect companies sold, installed and maintained telephone systems and competed directly with the local telephone company who continued to rent systems. You could purchase a system for $100,000 that replaced the system you were renting from the telephone company for $10,000 per month. From a financial perspective, the 10-month break-even point was a no-brainer. What businesses did not realize was that by buying their own telephone systems they were being pioneers. They often had a rocky road ahead.

It was around this time that I went to work for a consulting firm, back in New York City. Our specialty was showing businesses how to save a lot of money by purchasing their own telephone systems. We then proceeded to help them make the purchase and to manage the project. My job was project management. My Bell System background was what got me the job, but it did little to prepare me for overseeing the installation of large business telephone systems. I started learning from the very first day when I was directed to show up at a cutover (this refers to the process of changing over from the old to the new telephone system).

I learned at that time that the local Bell telephone company was not very helpful to customers who had left the fold to buy their own systems. The local Bell employees viewed the people installing the new telephones as having taken away part of the Bell job, which they had, so Bell was not too anxious to make cutovers a big success. Sometimes the interconnect system installers had just quit their jobs at the telephone company and were now on the other side from their former co-workers. You still needed the local telephone company to install new outside lines or to make adjustments to the lines already in place to accommodate the new system.

At that time, organizations were also required to install interface devices. These were circuit boards of questionable usefulness connected to the end of each Bell System outside telephone line to protect it from the foreign equipment that was being installed. These interfaces were rented to the customer by the Bell System company. This was a consolation prize awarded to Bell by the courts in exchange for having given up the lucrative system rental revenue. Actually, the interfaces just provided another point at which something could go wrong, which it often did.

The early systems did not always work as promised, so there were surprises. On several occasions, the installation company was unable to obtain the system that had been purchased so they put in a substitute system from another manufacturer, hoping that no one would notice.

I worked at a cutover where the customer had decided to purchase a used system. The installation company had not even bothered to remove the old coffee stains from the switchboard console! At this same cutover, it took a whole day before the installers could get the telephones to ring. When a call came into the switchboard, I would find out who it was for and run to tell that person to pick up their telephone handset as a caller was there waiting for them!

At another cutover, the customer had been told that the new telephones would be black, but the manufacturer only made white ones. That evening, the housings were removed from all the telephones and we spray-painted them black right there in the telephone equipment room.

At a typical cutover, a lot of people would be standing around looking worried, smoking cigarettes, drinking coffee and making periodic forays out of the telephone equipment room to reassure the customer: "Yes, at any moment now you will be back in business."

There was a lot of naivete during this time. This included the buyers, the consultants, and the interconnect companies. We were all making our way through uncharted territory.

The telephone system functions were clunky and did not always work. To successfully transfer a call without cutting the caller off was a major accomplishment. Just to keep this in perspective, note that the first telephone system purchased by a company was often the first system that enabled the system users to do anything by themselves. It is likely that they had never dialed another extension or transferred a call before. The systems they replaced were often cord boards where the switchboard operator had been responsible for handling all of these functions for them.

Those were the seventies. Several decades have passed and the process of installing a new business telephone system has smoothed out considerably. At a recent cutover, the CEO announced that the change to the new telephone system had been a "non-event" and the business operations continued without missing a beat.

Understanding the Telecommunications Industry

Who Does What

The purpose here is to provide you with an understanding of the different types of companies that make up the continually changing telecommunications industry.

Before 1984, the telecommunications industry was relatively simple to understand. Much of the United States was served by a single organization known as the Bell System and the balance of the country was served Independent Telephone Companies, several large and many small.

These companies sold most telecommunications equipment and services including (1) business telephone systems, (2) telephones, (3) cabling for your office, (4) local outside lines for placing both local and long distance calls and (5) data communications circuits with the needed equipment. Everything was billed to you on a single invoice each month.

After the break up of the Bell System monopoly and the resulting court judgments, companies in the telecommunications industry became more specialized. Some companies sold business telephone systems only. Some sold local outside lines and telephone calls. Others became long distance companies or specialized in data communications.

Now things are beginning to change again, and a new regulatory climate resulting from the Telecommunications Act of 1996 is permitting companies to expand into areas of the business from which they had formerly been prohibited. Companies have been merging or acquiring other companies with services that compete with or complement their own.

It is no longer possible to identify a company as purely a local telephone company or a long distance company. Instead, we classify companies by the types of equipment and services they sell and support. Keep in mind that many companies now fall into more than one of these categories.

Considering the pace of mergers, acquisitions, spin-offs and failures, nothing becomes out-of-date faster than telecommunications company names.

■ Telecommunications Network Services (Outside Lines And Calls)

● Local Telephone Lines and Local Calls

Local telephone lines come into your business, most typically on cables running in from the street to your office. They connect either your telephone system or individual telephones, faxes or modems to the outside world. They include dial-tone Lines also known as POTS Lines (POTS = plain old telephone service) or may be called auxiliary lines, combination trunks, direct inward dial trunks, T1s, and PRIs.

These lines are most typically delivered to your telephone equipment room or telephone closet where they are then physically connected to your telephone system or to a cable running to a separate telephone, fax or modem. These connections are made by your telephone system maintenance company.

To order these lines you call a Local Telephone Company. While there is now competition in many areas of the United States with more than one company selling local telephone lines and calls, you can call the Traditional Local Telephone Company; the one that has been around for years (Verizon or Bell South, for example). Since there is now competition, this company is sometimes called the ILEC (Incumbent Local Exchange Carrier) to distinguish it from the others. This company may have been part of the Bell System at one time or may be an Independent Telephone Company, depending upon where you are located. The competitors are called CLECs (pronounced "see-lex") (Competitive Local Exchange Carriers) or BLECs (Building Local Exchange Carriers). A BLEC is a variation on the CLEC that tries to corner the local telephone service market in certain office buildings where it has installed cabling and electronic equipment.

In some cases the CLEC is one of the companies historically recognized as a long distance company, such as AT&T. Most of the Long Distance Companies now compete with the Local Telephone Companies selling Local Telephone Lines and Calls in some geographic areas.

Most of the Long Distance companies got into the business of selling local telephone lines and calls by purchasing a CLEC.

There are often a number of CLECs that compete with the ILEC in certain geographic areas, mainly large cities where the opportunity is the greatest.

Some CLECs have run their own cabling into certain office buildings, but it is more common for them to rent the cabling that is already in place which is owned by the ILEC with whom they are competing.

These local telephone companies may also sell local telephone calls which are seldom detailed on your monthly bill, but instead billed as message units which have an associated cost. The calls are billed according to how far away you are calling and how long you speak. In some areas of the country, you may make an unlimited number of local calls for a fixed monthly cost known as a flat rate. Detailed descriptions of how local calls are billed by the ILEC in your area can be found in the front of the White Pages telephone directory they publish. CLECs tend to price local calls differently than the ILEC not only in terms of cost, but in terms of the initial period ,for which you pay even if you speak for just one second, and subsequent time period increments.

Many of the ILECs and CLECs also sell long distance calls and related services such as Calling Cards. They may also sell Cellular Service and Data Communications Services including circuits connecting you to the Internet.

It is likely that many companies selling these different types of telecommunications services were formerly separate companies that have been merged together. Therefore if you want to buy more than one service you may still feel as if you are dealing with separate companies. For example, there are often different sales people and different billing systems for each different type of service.

In some cases, these companies have separate arms that sell business telephone systems called PBXs. They may also sell a service historically known as Centrex which provides PBX capabilities using the functions of the local telephone company's own switching equipment at their central office rather than on the customer's premises.

Your local telephone company is also typically responsible for ensuring that your organization is appropriately listed in the white pages and yellow pages directories and in directory assistance. With the proliferation of competitors, these listings can no longer be taken for granted and must be regularly checked for accuracy and to be sure that callers can reach you when requesting directory assistance.

The local telephone company owns a number of central office switches (electronic equipment used to route telephone calls to their destination). The central office may be a large or small building depending upon the number of telephone service subscribers in the area. In large cities it is likely that you are located within a mile or two from this central office while in more rural areas you may be many miles away.

The cables that deliver the local telephone lines to your premises are run from this central office and may be routed underground or overhead on telephone poles. If you are in a high-rise building, the local telephone company may also own the riser cable that runs vertically from the basement and drops off telephone lines on each floor.

● Long Distance Telephone Calls and Point-to-Point Circuits

Long distance companies, also known as long distance carriers or IXCs (inter-exchange carriers) sell long distance telephone calls. When you pick up the telephone in your office and place a long distance call it may be sent over an outside line connecting you directly to your long distance company. The call may also be sent over the outside lines connecting you to your local telephone company, who then passes the call along to the long distance company you have told them to use on your behalf. This is called your carrier 'pic'. The way you have your telephone system set up will control which of those two routes your call takes. This routing is controlled by the Automatic Route Selection capability of your PBX.

Long distance companies also rent circuits that permanently connect two or more offices of a large organization. These circuits may carry voice, data or video signals. The circuits are sometimes called leased lines or dedicated lines.

There are some large long distance companies and many smaller ones that cover the entire United States or a specific geographic area. Most handle international calls as well. U.S. companies may have relationships with carriers for local and long distance calls in other countries, since a call may be carried on circuits of several different companies before reaching its final destination.

Even in the U.S., a long distance telephone call is most likely to use the circuits of the traditional local telephone companies at both ends of the call. The long distance companies pay the local companies for the use of these circuits.

If you're renting a circuit from a long distance carrier connecting two of your offices, the local telephone companies at both ends may provide the local leg or local loop (mean the same thing) of the circuit, even though you are billed by the long distance company.

Sometimes long distance companies rent circuits from each other. Some calls may still be transmitted by bouncing the signals off of communications satellites. Transponders on the satellite may be shared by different long distance companies. The trend is away from satellite communications as the use of fiber optic cable becomes more widespread and its resulting capacity to handle call volumes increases. However there are still remote areas relying on satellite and microwave radio communications.

Long distance companies generally refer to their network, which is the collective group of all circuits over which calls are sent or permanent point-to-point connections are made. The network may include a variety of transmission media including copper cable, fiber optic cable (underground and above ground), and to a limited extent, microwave and satellite communications. The network also includes electronic hardware to provide the media with the capability to transmit.

Long distance carrier networks have electronic switching equipment for connecting and routing calls, enabling a large number of people to access a common group of outside lines or circuits. These long distance company switches are located at what is sometimes referred to as the carrier POP, meaning point-of-presence or the NOC (network operating center). You may reduce your costs by having a circuit that connects your telephone system directly to this POP. The circuit may be rented from the local telephone company by the long distance company, although you will probably be billed for it by the long distance company. The cost of this circuit is called an access charge.

Smaller companies and residences reach the long distance carrier networks in most U.S. locations through equal access. Each of the major long distance companies in your area rents space for its own electronic switching equipment at the local telephone company central office. The programming in the central office switch will route your call over the network of the long distance company you have selected.

There are many companies selling long distance calls, representing a variety of long distance companies. You can buy long distance calls directly from the carrier or from one of these companies which may be called a reseller, rebiller, aggregator or a variation on the same concept. These companies are providing bulk buying, representing large numbers of customers, and are therefore able to negotiate lower rates for the organizations in their group.

Since a long distance call has become a commodity in that most calls are of equal quality, not counting cellular service which can vary, one must select a long distance company based upon reputation, service and clarity and accuracy of billing. The cost of a long distance call is based upon many variables. These include distance, time of day, duration, total number of calls made (higher volume = lower costs) and special promotions in effect at the time you select a long distance carrier.

Long distance companies provide 800 numbers (may also be 888 or another prefix) enabling callers to reach you without paying for the call. You pay for it instead. This is also called a toll-free number. Along with the 800 numbers, you can buy some sophistication in the routing of your calls. For example, you can have a single 800 number with callers from each state routed automatically to the nearest branch of your company when they dial the number. Your 800 number can also be programmed to send your calls to your New York office until it closes. After 5PM EST, it will reroute the calls to your California office.

You can receive reporting of the telephone numbers of the people who are calling your 800 number. This is called ANI (pronounced "Annie") or automatic number identification. More new services are being introduced regularly.

Long distance companies also provide: 900 services, 700 services, credit cards, toll fraud prevention programs, and customized bills and management reports on paper, on CD-ROM or on-line Web access and other related services.

The Telecommunications Act of 1996 has enabled the traditional long distance companies to compete for local telephone service. The local telephone companies, meeting certain criteria, can now sell long distance service as well.

Most major long distance companies also sell local service, Internet access and cellular service.

Calling Cards

A calling card enables you to place calls from any telephone using your unique calling card number and personal identification number (PIN). Calling cards are provided by both long distance companies and local telephone companies. There actually is a small wallet-sized plastic card (like a credit card) that is issued to each calling card holder. In some public telephones including those in airport travelers' lounges, this card can be swiped through a slot on the telephone which reads the calling card number from the magnetic strip on the back of the card. If the telephone is not equipped with this slot or if the card holder chooses not to carry the card around with him, a calling card call can be placed by dialing a toll free telephone number of the service provider (11 digits), the telephone number you are calling (11 digits) then the calling card number (10 digits) and PIN (4 digits.) Despite the inconvenience of dialing 36 digits, the use of calling cards is widespread for business travelers, although decreasing as the use of cellular telephones increases.

There are still some calling cards that require fewer digits to be dialed by the user in that there is only a need to dial "O" plus the called telephone number before being prompted to dial in the calling card number. Typically these are the cards issued by the traditional local telephone company or the original long distance telephone company, before there was competition.

This type of call is an example of what is called a zero plus call.

Most calling card calls are charged not only a cost per minute, but a surcharge on a per call basis. These charge vary dramatically depending upon the rates you have negotiated with your particular service provider.

This is sometimes called a "bong" charge after the sound that is sometimes heard on the telephone line by the caller before dialing in his calling card number.

Calling cards calls are billed on your telephone bill and are identified by the calling card number placing the call.

● Pre-Paid Calling Cards

Another type of calling card is a pre-paid calling card. This also is a wallet sized plastic card, but unlike the other type of calling cards, you pay in advance for a pre-set number of minutes of calling. You can buy these cards in many retail locations like newspaper stands and drugstores.

The pre-paid calling card companies buy the service from a local or long distance service provider. Some local and long distance companies also sell the pre-paid cards themselves.

Data Communications Circuits and Services

Most telecommunications service providers sell both voice and data communications circuits and services, although some are more focused on data. Others may have separate groups of salespeople to handle the more complex and often more profitable sale of data services.

Data communications is a broad term that refers to any type of computer-to-computer communications.

Many large organizations with multiple sites set up a data communications network using circuits and related services and equipment from the telecommunications service provider.

Data communications services include individual circuits of different speeds or bandwidth, having to do with the throughput capability. Data Communications also refers to networks connecting multiple sites.

Data communications also includes an organization gaining access to the Internet and using its capabilities to communicate among multiple locations.

Some telecommunications service providers set up what is known as a VPN (virtual private network) or intranet which uses Internet-like capabilities on the service provider's own network to set up data communications for its customers.

Once a data communications circuit or entire network is set up, it is technically possible to use some of the capacity to transmit voice communications as well. Some organizations do this and others are experimenting with it as a means of cost reduction.

There are many quality of service considerations and associated start up costs for using a data communications network to transmit voice.

Cellular Telephone Service

Most telecommunications service providers now sell cellular telephone service. While cellular telephone calls often use the existing network of the telecommunications service provider for some portion of the call (may be called land lines or wire based services), the initial connection to the network from the cell phone is made without a physical connection based upon a cable.

Cellular service providers erect transmission towers at different points within a geographic area to enable the cellular transmissions.

ISPs

The ISP or Internet service provider provides access for organizations to reach the Internet. They may also provide a group of Internet related services such as the hosting of an organization's e-mail or web site. The ISP is reached either by the computer dialing into a certain telephone number or over a permanent connection provided by the ISP using the cable of a local telephone company.

Cable Television Companies

Cable Television companies are making some inroads into providing services traditionally offered by local telephone companies, such as a local telephone lines for placing and receiving calls and data communications including access to the Internet. While the cable TV company has cable running into most homes, it still needs to invest in considerable amounts of equipment to turn its system into one that can be used for telephone service and Internet access. Since cable TV does have much more of a presence in homes than in businesses, cable TV companies are selling telecommunications services primarily to residential users.

Pagers and Messaging Devices

Some wireless companies also sell the traditional beepers and devices that provide much more advanced ability to receive and send text messages. Some devices now enable the receipt and sending of e-mail as well as Internet access.

Since some cellular telephones are equipped to handle the text messaging and e-mail as well, the lines between pager companies and other wireless service providers are blurring.

Telecommunications Systems (Equipment)

Business Telephone Systems
(PBX and Key Systems)
Installation and Maintenance Companies

The telephone system installation and maintenance companies in your area can be found in the yellow pages under Telephones.

Sometimes these companies are called telephone system vendors. They used to be called interconnect companies although this term is now rarely heard. These companies developed starting in the 1970's when it first became legal to purchase a telephone system from an outside supplier rather than renting it from the local telephone company.

Some telephone systems are installed and maintained by the same company that manufactures them. These installation arms of the manufacturers are often former interconnect companies who first sold, installed and maintained the products of the manufacturer and were subsequently acquired. Other installation and maintenance companies are authorized distributors and carry the product line of one or more telephone system manufacturers. In either case, the services you can expect from a telephone installation and maintenance company include the following:

▶ You can purchase a business telephone system from them (including the backroom equipment and the telephones).

▶ They will install the system. Installation includes pulling the cable in the walls; connecting the telephones to the cables; and connecting the control cabinet of the telephone system to the other ends of the cables and to the outside lines brought in from the local telephone and long distance company. They will also install the switchboard console, used as a central answering point.

▶ They may also install cable for your computer network. It does not make sense to have two separate companies (one for telephones and one for computers) running cable back to a central point since a significant portion of the cost is for labor.

▶ They will program the system in a manner to complement your organization's particular way of operating. Programming the telephone system for the way in which it will be used is as important as buying the right system. Most telephone systems have a set of rules determining how they may be programmed. Programming determines such things as which extensions are picked up by which telephones and what happens to a call when it rings on an extension which is unanswered or busy. The interaction between the telephone system and the voice mail system is a key element of system programming and how successfully your system handles callers.

▶ They will train your staff how to use the telephone system and provide an instruction manual.

▶ They will handle changes to your system such as installing new telephones, rearranging telephones and making programming changes. If you wish, someone on your staff can learn to make some of the system program changes with a Maintenance Administration Terminal or MAT. The changes made with the MAT are often called MAC work which stands for moves and changes.

▶ They will handle all repairs to the telephone system and the cabling.

▶ They may act on your behalf to interface with the local telephone company. If there is a problem with one of your outside lines, they will report the problem and follow up until it is resolved.

▶ They sell other related systems such as Voice Mail, Automated Attendant and Call Accounting which they will install and maintain.

▶ They may also be authorized representatives for a local telephone company or a long distance carrier, from whom they receive a commission for selling certain types of services. This is called being an *Authorized Agent* or a *Reseller.*

The telephone installation and maintenance company is not the place to call for problems with telephone bills or for help in deciding the types of local and long distance telephone services you need.

Some companies can provide service nationally in the U.S., but most serve smaller areas. When evaluating a company, it is important to focus on the support available in the local area where the telephone system will be installed.

Some smaller telephone systems (for less than 20 people) can be purchased in telephone stores or through catalogs. It's a good idea to have an experienced telephone installer put the system in for you rather than doing it yourself, unless you want to learn by doing.

● Business Telephone System Manufacturers

There are approximately ten major manufacturers of telephone systems for large organizations and perhaps another twenty to thirty that make smaller size systems. Two of the manufactures (Avaya and Northern Telecom) evenly share approximately 50%

of the business telephone system marketplace worldwide. Avaya was formerly Lucent and before that it was AT&T.

These companies manufacture the control equipment (PBX cabinet, circuit boards, processors, etc.) and the telephones themselves that work only with the system from that particular manufacturer.

Local or Long Distance Company Switching Manufacturers

The major companies that manufacture telephone systems (called PBXs or switches) for large businesses also manufacture switches for use by either a local telephone company or a long distance company to connect the many calls from all sources to destinations passing through their network. These are often called central office switches and are said to be part of the public switched telephone network or PSTN. PSTN is often used to refer to the separate network other than the Internet. The network consists not only of these switches but also of the connections between them, most often made via copper or fiber optic cable.

Manufacturers of Telecommunications Hardware
Used by the Local and Long Distance Telecommunications Service Providers in their Networks

Some of the telephone system manufacturers also make other equipment used by local and long distance service providers in their networks. These may be called bridges, routers, data switches, smart modems, hubs, etc. These facilitate the transmission and proper routing (sending to the right destination) of voice, data and video transmissions.

Some companies that were originally only in this business are trying to move into the telephone system business by manufacturing their own versions and visions of next generation telephone systems.

Data Communications Hardware

Again, there is overlapping in what different companies make. What we are calling data communications hardware are the electronic devices primarily used on an organization's premises in support of communications among a group of computers either at a single site or connecting multiple sites.

This is often the same type of equipment used by the telecommunications service providers themselves in their own networks such as routers, bridges, data switches, etc.

Specialized Business Telephone System Manufacturers

TRADING TURRET COMPANIES

There is a very small group of companies who manufacture a type of telephone system called a trading turret system, sometimes called a dealing system, particularly in Europe. These are large multi-button telephones, usually 60 or 120 buttons, somewhat like a key system, but designed with the brokerage trader in mind. Traders require instantaneous communication with a large number of other traders. This is accomplished through point-to-point circuits that appear on the buttons of the trading turret.

The newer versions of these systems are more dependent upon software which provides the appearance of a multi-button telephone on a computer screen. This enables access to many pages of different outside lines through which the trader can scroll.

As with telephone system manufacturers, some companies have installation and maintenance arms while others sell their products through distributors who install and maintain the systems.

AUTOMATIC CALL DISTRIBUTION SYSTEMS FOR CALL CENTERS (ACDs)

Several companies manufacture telephone systems specifically designed to handle large call centers, such as airline reservation operations, groups of order takers handling catalog sales or help desk/customer service centers. These systems are specialized switches designed to handle high volumes of calls. They route the calls to different groups of agents or representatives and provide management statistics. These statistics include how many calls are handled by each person, how long callers wait to be answered, and how many callers are on hold at any given time.

Most PBXs can be set up to work as an ACD.

Many applications linking computer and telephones are being set up in conjunction with an automatic call distribution system

▶ PERIPHERAL SYSTEM INSTALLATION AND MAINTENANCE COMPANIES

While many telephone installation companies also sell voice mail, automated attendant, call accounting and facilities management systems, another small group of companies sells only these systems without actually selling the telephone system itself. The advantage of buying from these companies is that they tend to have more in-depth knowledge of the types of systems they sell than the telephone installation and maintenance companies do. Also, the systems that they sell are often more technologically advanced than the systems sold by the telephone installation and maintenance companies.

The disadvantage of working with such a company is that you now have an additional supplier to manage. If you are buying from one of these companies selling peripheral equipment only, be sure that the systems are already working with the type of telephone system with which you plan to use them. Also, be sure that this company has worked with the telephone system installation and maintenance company you are using or is willing to do so, so that you're not left to do a lot of coordination between them during and after the system installation.

▶ PERIPHERAL SYSTEM MANUFACTURERS

While some business telephone system manufacturers also make peripheral systems, other companies specialize in the peripheral systems only. Some hardware and software and others developing the software only. These include:

a) Voice mail and automated attendant companies who may also sell software applications related to these.

b) Interactive voice response companies (IVR) who manufacture but also provide the development of IVR solutions (such as being able to check your bank balances by using a touchtone telephone.)

c) Fax Server Companies

d) Call Accounting, Charge-back and Facilities Management Companies

Manufacturers of call accounting systems tend to be small organizations with under $10 million per year in business. There are one or two that were acquired by larger companies, but still tend to operate as separate entities.

This type of company makes and supports the call accounting software and other related software packages for things such as facilities management or record keeping and telecommunications work order processing.

These companies typically prefer to provide the computers and related equipment, such as printers, with which their software operates. Increasingly their customers are providing the hardware themselves and may be using the systems in a networked environment meaning that multiple people within the same office or connected offices can access the system to make changes or view information.

Recognizing the work involved in the administration of on-site Call Accounting Systems, some organizations are opting to use the services of Call Accounting Service Bureaus, discussed later in this tutorial, to accomplish the same objectives.

▶ CABLING INSTALLATION COMPANIES

Your telephone installation and maintenance company will install the cable for both the telephone system and the computer network.

Other companies are in the business of running cable only. The company from whom you are buying the telephone system is likely to sub-contract the cable pulling to one of these companies anyway.

These companies are often electrical contractors. When using electrical contractors for telecommunications cabling, it is important to be sure that they have experience installing this type of cabling.

If a separate company is installing the cable, be sure that they will certify the work and that the company installing the telephone system (or computer system) on that cable will accept it. The telephone installation company putting in a new telephone system may wish to charge extra for testing the cable that they did not install, sometimes called toning and testing. Certification of cabling typically adds five to ten percent to the price, but it's a good insurance policy. There are different levels of certification, so it's important to be very clear on what is being requested and what happens when something goes wrong in the future.

No matter how small the project, it's a good idea to put cabling specifications in writing and obtain competitive bids. You may wish to take a bid from the telecommunications installation and maintenance company and one from an electrical contractor to see how they compare.

An improperly installed cabling job can create problems with systems that may never be resolved.

▶ MANUFACTURERS OF CABLING AND RELATED HARDWARE

While people often think of cabling and the hardware associated with it to be somewhat of a commodity, there is actually considerable difference in quality among the various manufacturers. Different companies make copper cable and the associated hardware like jacks and distribution panels while others manufacture fiber optic cable and the associated electronics.

▶ DISTRIBUTORS OF CABLING AND RELATED HARDWARE

These companies typically represent multiple manufacturers and sell their cabling and hardware to companies who install the cable.

▶ VARs (Value Added Resellers)

Companies traditionally specializing in installing office computers and computer networks are now selling business telephone systems and peripheral systems, as well. Their intent is not so much to get into the telephone system business, but rather to enable their customers to experiment with applications that use capabilities of both the telephones and the computer networks, such as having a screen of customer information pop up on the computer with the arrival of a call from that customer called a screen pop.

▶ Intercom Companies

The office telephone system did not always provide the capability for calling others in the same office, now known as internal calls. There was once a larger group of companies that sold intercoms for inter-office communications. Several of these companies still remain since there are some environments like a Wall Street brokerage trading room, that demand instantaneous voice communications with other individuals in the same office or same large room.

The intercoms are separate desktop devices with their own control equipment for the backroom and using their own separate pairs of wires on the office cabling infrastructure.

▶ Wireless Telephones within the Office

As some office workers move around the workplace for most of the day, there is an increasing need for them to be able to continue to receive their calls.

Some manufacturers of telecommunications systems also sell wireless telephones or wireless headsets to work with that system, using base stations, with the same concept as many wireless telephones used in the home. These tend to be comparatively high priced and work within a limited area.

Several companies manufacture separate wireless systems for the office environment requiring the installation of transmitters in the ceiling enabling coverage of the entire office. These systems are separate from the office telephone system but can be set up to work in conjunction with it, enabling workers to receive calls to their desktop telephone from anywhere in the office. The drawback of these systems is that they tend to cost as much as the telephone system itself.

Still another approach to wireless communications within the office is to enable workers to use their cellular telephones as a "wireless telephones working with the office telephone system" when they are in the office. This requires an interface with the office PBX and the cellular system that often takes place within a communications server in the office.

▶ OVERHEAD PAGING

While fewer conventional offices are using overhead paging as a means of locating people within the office, there are still certain environments, such as warehouses and manufacturing facilities, where overhead paging is appropriate and useful.

A well-designed overhead paging system requires the knowledge of an acoustical engineer to properly space and regulate the speakers.

While Telephone Installation and Maintenance companies can and often do install and maintain an overhead paging system, engaging an engineer to design the system is a worthwhile investment to enable the system to work properly and preserve your hearing!

The paging system typically consists of an Amplifier located in the telecommunications equipment room, and cabling connecting the amplifier to ceiling speakers.

The office telephone system can be programmed to enable some or all users to access the paging system, typically by dialing an access code such as "8". Sometimes paging capability is limited to the switchboard attendant only.

Some smaller business telephone systems have paging capability built in. The page comes out of the same speaker on the telephone as the intercom. The drawback to this type of paging is that it is usually limited to about 15 telephones and may be distracting since it is heard on every desktop. (It can be eliminated from certain telephones if necessary.)

Software And Service Companies

Software Developers

There are many companies developing a wide variety of software applications to be used in conjunction with the telephone system. Many of these come and go without ever taking hold in any significant way. Others tend to evolve as their development proceeds. Most require considerably more customized programming that is apparent at first.

Many of these applications are developed for the incoming call center environment where large numbers of calls are handled for customer service or sales.

Some of the types of software include:

▶ Workforce Management, helping Call Center Managers to staff properly

▶ CRM or Customer Relationship Management, enabling organizations to create and maintain strong relationships with their customers through personalizing their communications with them.

▶ Convergence Applications enabling users to merge voice and data to, for example, handle voice communications and e-mail over the same desktop or wireless device.

Call Accounting and Charge Back Service Bureaus

With multiple telecommunications companies providing a variety of systems and services and with expenses on the rise, the need to manage telecommunications within the organization is increasing. There is a small group of companies who write and support software to assist organizations with the management of telecommunications assets and often computer assets as well.

These software systems keep track of the system configurations (circuit boards and spare capacity), the actual desktop devices such as telephones and PCs, the cabling, and the circuits from both the local telephone company and the long distance companies. In addition, they may track work orders so that as changes are made to the system, all information is automatically updated. Company telephone directories are also generated from the software.

They provide cost allocation and charge back capabilities for telecommunications equipment, services and calls.

● Directory Assistance

A number of different companies using different databases now provide directory assistance. As a result, the likelihood that you will get accurate information from directory assistance is diminishing. This is the service you may reach by dialing 411 to reach your local telephone company or the area code plus 555-1212 to reach your long distance company. The call centers where the directory assistance operators are located may not be anywhere near your city or be familiar with it.

● Directory Publishers

Most local telephone companies still publish some form of the traditional white and yellow pages paper directories. The publishing may be done by a separate company, however, often formerly part of the local telephone company. There are also competing companies offering directories specific to a particular city or geographic region.

These very thick paper directories now compete with Internet based directories, but it appears that they will be around for the foreseeable future. The yellow pages represents significant revenue in advertising for which the customers pay every month while the directory is in effect.

Professional Services

Telecommunications Consulting

These companies offer a range of professional services based upon hourly or project fees. Since no two are exactly alike, it is important to identify the services and capabilities of each company. These services may include:

a) Assessment of your organization's use of telecommunications technology.

b) Assistance with the development of requirements, evaluation of alternatives and procurement of telecommunications equipment and services.

c) Telecommunications Contract Negotiation

d) Telecommunications Project Management

e) Telecommunications Expense Review and Cost Reduction

Telecommunications Bill Auditing

These companies find errors on your telecommunications bills and are paid a percentage of the refund or savings they arrange on your behalf. They determine whether your service providers are complying with your contracts and the related telecommunications tariffs. They may also determine whether all of the services for which you are being charged are actually in place and working.

Telecommunications Bill Management

As the total number and complexity of telecommunications bills increases, organizations are looking for outside help to manage them. This breed of company is still developing and may offer services such as monthly review and validation of bills, tracking of expenses and correction of billing problems. Some larger Bill Management Companies also pay all of your telecommunications suppliers and invoice you with a single master bill.

Telecommunications Outsourcers

Although the concept has been around a long time, the idea of outsourcing a variety of telecommunications management and maintenance functions is still alive and well, although the quality of telecommunications outsourcing has historically left much to be desired. The following descriptions provide you with some ideas of the types of outsourcing available:

▶ *Complete Outsourcer.* If you completely outsource your organization's telephone system and services, this means that you have delegated all responsibility to an outside firm including the repair and maintenance of your PBX, provision of local and long distance service, reviewing and payment of telephone bills and a variety of other services that should be carefully spelled out in your contract. The outsourcer is paid a monthly fee and also makes money reselling you equipment and local and long distance service. This type of a relationship may seem like "the fox minding the hen house," since the organization doing the billing is also responsible for approving the bills. Nevertheless, it is working well for a number or large organizations.

▶ *Telecommunications Management Outsourcer.* This type of outsourcing manages your telephone system and service providers, but the company does not sell you equipment, local or long distance service as a Complete Outsourcer may. The Telecom Management Outsourcer is the equivalent of having an in-house telecommunications department, but an outside firm staffs it.

▶ *Out-tasker.* An out-tasker, either a firm or an individual, takes the complete responsibility for one or more specific tasks. For example, you may out-task the updating and production of your telephone directory. Another example would be to out-task the work order processing and record keeping for changes to your telephone system.

▶ *Consultant.* Many people refer to all subcontractors as consultants, so this is one view of the term. In the stricter sense, a consultant is an individual who is charged with analyzing a problem or assessing a situation and providing a recommendation. A consultant may also execute steps necessary to carry out the recommendations. A consultant may be an individual, may have a small company or may be a member of a small or large consulting firm.

▶ *Technician.* Like "consultant", this is a term that has varying definitions. A technician is typically someone who can do the things a purely administrative person cannot. For example a telephone system administrator can write up requests for changes to the telephone system such as adding two new extensions to a particular telephone. The administrator, if trained to do so, can also make program changes to the telephone system to add the two new extensions. If, however, a new circuit board for the PBX was required for those new extensions, or if any rewiring had to be done, that would be the job of the technician.

▶ *Subcontractor or Independent Contractor.* These term are used interchangeably and typically refer to an individual who is providing work to you for a fee, but is not on your payroll. The fee is usually per hour or per day (also called "per diem"). Depending upon expertise or experience, this person can provide support in any area of your department. Sometimes an entire firm is referred to as a subcontractor. Any of the above categories may be considered to be a subcontractor or independent contractor.

◻ Outsourcing Resources – Who To Call

▶ *Telecommunications Outsourcers.* Some firms identify themselves as Outsourcers. At present, they seem to be going after the larger, multi-site businesses. Initially they outsourced the

Information Technology systems, but have added Telecommunications to their capabilities.

▶ *Large Telecommunications Vendors.* Some large telecommunications installation and maintenance companies or local and long distance network service providers will take on a complete outsourcing of telecommunications equipment and services.

▶ *Consulting Firms.* Some consulting firms also provide ongoing support which may include Telecommunications Management Outsourcing or Out-Tasking in specific areas of expertise. The consulting firm will assess your requirements and may provide management of the Out-Tasking, which may make it a better choice than simply hiring a group of subcontractors from separate firms or from a temporary agency.

▶ *Specialized Support Firms.* Some firms provide support specific to a certain type of product or service or, in some cases, specializing in services relating to equipment from a particular telecommunications vendor.

▶ *Temporary Agencies Specializing in Telecommunications Staffing.* These firms operate like any temporary agency, but maintain resumes on people with different capabilities in the telecommunications arena. Try to find a firm that is truly a temporary staffing firm, rather than a company whose specialty is really permanent placement and only dabbles in the temporary market.

This concludes our explanation of the Telecommunications Industry. It is varied and continually changing as the forces of both governments and the marketplace continue to shape it.

Purchasing Telecommunications Equipment and Services

*You don't really enjoy
tearing up dollar bills —
do you?*

Purchasing Local Telephone Service

In the beginning there was the dialtone...

Here are some ideas to help you to purchase and negotiate for local telephone service.

By local telephone service we mean the physical outside lines that connect your telephone system to your local service provider and the local calls themselves.

Historical Perspective

Traditionally, everyone has been buying local telephone service from the "telephone company," a public utilities commission regulated division of one of the RBOCs (Regional Bell Operating Companies) formed in 1984 when the old Bell system was split up. These divisions are now what is known as "Incumbent Local Exchange Carriers" (ILECs). Although these companies still have the lion's share of local business, the Telecommunications Act of 1996 changed the rules enabling many new competitors to enter the marketplace, including the long distance companies.

In the late 1980's – early 1990's there were a few competitors to the ILECs, most notably Metropolitan Fiber Systems (also called MFS, now part of MCI/Worldcom) and Teleport Communications Group (also known as TCG and now part of AT&T known as ALS or AT&T Local Service).

Who Is Selling Local Service Now?

There are several broad categories of companies who sell local telephone service. This varies depending upon your geographic location, with more options available in areas with a high density of businesses with increasing demands for service:

1. The traditional local telephone company.

2. Long distance companies that have now acquired local service capability, including MCI/Worldcom and AT&T.

3. Competitive Local Exchange Carriers (also known as CLECs) are the newest competitors. (As of 2001, many of these companies are not doing well, so check them out carefully.) They may provide the service on their own transmission facilities or rent facilities of others and resell them to you. They may also combine the two approaches.

4. Building Local Exchange Carriers (also known as BLECs) who have a presence of some type directly in your office building and may rent space from your landlord or have a revenue sharing arrangement with them.

5. Cable TV Companies (buying local telephone service from cable TV companies is available in some areas, but relatively new and not widely offered as an option for business in most cities.)

6. Resellers (some are called agents), who represent one or more of the above companies and probably sell long distance service and Internet access as well.

What Type Of Outside Lines Do You Need?

The most frequent mistake organizations make in purchasing local telephone service is buying more than they will ever need. If you currently have a telephone system, you can request a traffic study from the company who supports your PBX. The traffic study monitors the lines on which you place local calls for a week to determine how many outside lines are in use at the busiest times of day. If you have 40 outside lines but there are never more than 20 in use, you can safely disconnect some without affecting service.

Many organizations now have two types of outside lines for local service (1) Direct Inward Dial lines for incoming calls and (2) Both-way lines that are used for some incoming and all outgoing calls. In the past, each of these lines was delivered to your telephone equipment room on a separate pair of copper wires, but now more typically they will be delivered on what is called a T1 (24 outside lines on 2 pairs of wires) or a PRI (like a T1 but providing more flexibility to handle both incoming and outgoing calls within its capacity) if the size of your organization warrants this.

While the ILECs (incumbents) are just beginning to sell long distance service, if you are buying from one of the CLECs (competitors), look into combining access to local and long distance service on the same T1 circuit. If you don't need the entire capacity of the T1 you may also use part of it for Internet access, but you must set this up when you purchase the service and must also install the appropriate electronic equipment (buy it or rent it) to make it work.

Buying Local Telephone Calls

The cost of a local telephone call has not yet come under the same scrutiny by the general public as the cost of a long distance call, but the comparisons are similar. Typically there is an "initial period" for which you are billed regardless of the length of the call. With the ILEC this may be three or five minutes with an additional cost per minute after that, likely to be billed in one minute increments. CLECs may have shorter initial periods and shorter increments for additional time. Different initial periods and increments may apply on calls to different destinations. Try for six second increments at a maximum and an initial period of no more than 18 seconds. This will lower your overall cost particularly if you make many short calls.

In some areas of the United States, businesses can still get a "flat rate" meaning unlimited calls to the local area. CLECs sometimes call this "free" home region calling, but the cost of the outside lines is typically higher to offset the cost of handling the calls.

While CLECs compete with ILECs for calls to the local area, the traditional long distance companies add to the competition for what are called intra-LATA calls (LATA = local access transport area). This is a geographic area outside your local calling area. So get pricing from the ILEC, the CLEC and a long distance company to do a complete comparison on the intra-LATA rates.

In order to compare rates from the traditional telephone company, the CLEC and the long distance company, it is necessary to compile a profile of your current usage in terms of how many calls, how many minutes, the average length of a call and the destination of the calls. In you have a Call Accounting system on site or use a service bureau, such a report can be obtained. Obtain it for a representative month, not the Christmas or summer holidays.

You may also ask your ILEC or CLEC to compile this for you, but make sure you validate the source and accuracy of the information yourself. They may charge you for this service.

It is not likely that the ILEC will be able to negotiate the cost of local or intra-LATA calls since they are bound by tariffs as a regulated entity (So are the CLECs, but they can be more flexible than the incumbents since they have more options in how they buy the components of the services they deliver.) If you have a sufficient call volume, the ILEC or CLEC can negotiate a local service discount. The CLECs will typically offer a lower rate than the ILEC. If they are reselling the service of the ILEC it may not be dramatically lower, but if they have their own network they may save you up to 50%. Do your comparisons carefully before making a change. Any change can cause disruption in service and needs careful planning.

In addition:

▶ Investigate the CLECs track record, financial stability and references.

▷ Ask the references how smooth the process was to convert to the new local service provider.

▷ Find out who will be responsible for being sure your correct directory listings provided to all sources your callers may reach.

▷ Identify the frequency and length of any recent service outages in your area.

▷ Find out what the bills will look like and from whom they will be sent? Can you get good management information along with the bills for the number, duration and destination of calls? How easily can you validate that the correct discounts have been applied? Is the billing available on electronic media (CD-ROM) or through a web site?

▷ Find out who you will call for service problems? Do you call the reseller if you're using one? If you are buying local and long distance service from the same company, can you deal with just one representative? Does your account executive have the knowledge to explain bills and the clout to correct service problems or do you need separate relationships with the billing and service departments?

▷ Since things are changing, don't commit to any long term agreements (2 years at most). The ILECs have many customers locked into 120 month agreements ("Hey! That didn't sound like it was 10 years when I signed it!"). Also negotiate how you will get out of the contract if you move or if your business drops off.

How Is The Service Delivered To Your Premises?

You will be in a stronger point to negotiate price if you understand how the service is delivered and who owns what part of each element in the delivery. In general, a local service provider (CLEC) can negotiate more on the parts of service delivery over which they have the most control (usually the parts they own).

There is a new acronym called an UNE (pronounced ooney – like looney without the "l"). This stands for unbundled network element and refers to the ILEC breaking up the elements of delivering local telephone service so that the CLECs can buy the pieces they need to compete with the ILEC! (This from the Telecommunications Act of 1996.)

So find out who owns what in getting the service to your premise including (we'll start from your premise and work backwards.)

1. Electronics in your own telephone equipment room. This may be some type of multiplexer or what is called a CSU/DSU needed to work with a T1.

2. Horizontal cable from your own telephone equipment room to the telephone closet outside your premise (to the riser closet if you are in a multi-story building.)

3. Riser cable to the basement of your building. (Some CLECs and most BLECs run their own riser, others rent the needed pairs of wires from the traditional telephone company or perhaps from the landlord who has installed his own cable and controls it.)

4. Electronic equipment in the building, but not on your premises. Most BLECs and some CLECs locate switching equipment in the basement of your building through which your local lines are provided.

5. Connection from the building back to the off premise switching equipment. Find out the capacity of the circuits the local service provider has run into the building (T1, T3, PRI, etc.) and whether you are sharing it with other building tenants. Also does the CLEC rent the copper or fiber cable from the ILEC, from another transport provider (also known as a carrier's carrier), or do they have their own cable facilities run into the building and back under the street to the off premise

switching equipment? The service may also be brought into the building via a roof based microwave dish. You still need to ask the question about the capacity and how many tenants are sharing what size of circuits coming to the building.

6. Who owns the switching equipment from which you receive dial tone on your outside lines? Where is it located? (And what happens if it crashes?). Is it located in a "carrier hotel" with other CLECs? Is it co-located at the ILEC's premise? How are your calls handed off to your selected long distance carrier?

Another consideration if you have a group of direct dial telephone numbers that need to be "ported" from your ILEC to the CLEC: it is important to understand the process before it gets underway to ensure that it will be smooth.

As a final consideration, you will need the support of your telephone system service company to make the transition – taking the old lines out of your telephone system and installing and programming the new ones.

As we mentioned above, buying local and long distance telephone service, Internet access and cell phones and pagers, too, may all be done with a single company. But in all cases, it pays to shop around to ensure the best prices and a relationship with a company who really understands the services they are delivering.

Ordering and Provisioning Telecommunications Services

To be patient is a virtue.
To be detail-oriented is divine!

A great amount of energy goes into selecting the correct telecommunications equipment whether it is an entire PBX (telephone system) or a simple modem. An equally important process is the ordering and provisioning of the telecommunications circuits that connect your equipment to the outside world. Here are some suggestions to help you.

"The Rules"

First, the rules that apply to all orders, large or small.

▶ Start a file for each order you place. Keep all contact names and telephone numbers, correspondence and a log of conversations (names, dates, details of conversation) within the file.

▶ Make sure you are speaking to a person who can take your order and understands the service. Getting to the right person can be a challenge.

▶ Start by getting the complete name, company name, address, telephone number and fax number of the person taking the order. Write it all down and tell them you will be faxing a confirmation letter. (Even if you send an e-mail, send a fax, too and follow up to be sure it is received. The fax is not as easy to overlook.)

▶ If the first person you speak to says someone else will call you back to take the order, insist on getting that person's name, department and direct telephone number. Ask for a specific time commitment for the call back. Keep the name of the original person so you can call them if the promised call back does not materialize (a frequent occurrence!).

▶ When you place the order, ask for the order number(s), circuit number(s), installation charges, monthly charges and due date. This information will not be available when you place the order, but can be provided on a call back. Wait until you have all of this information before writing and faxing your confirmation letter. If someone will be connecting equipment to this circuit for you, send that person a copy of the letter. If there are several weeks or months between the time you place the order and the due date, call several times during that interval to confirm the order.

▶ Ask for a name, department and telephone number of a person you can call as the due date approaches and on the due date to confirm that the order in "in the field" and that the appointment will be kept. (Appointments are frequently missed.)

▶ Ask if there are any physical requirements for the installation of the circuit. For example, if it is being installed in a new space where construction is underway, it is important that there be access from the telephone closet (typically in a common area of the building outside your space) into your own telecommunications equipment room. Some building codes may require that a pipe be installed connecting your room to the building telephone closet. The cable is then run inside the pipe. Within your equipment room a wooden backboard must be mounted on the wall to accept the demarc (point of demarcation – usually some type of jack on which the circuits are terminated). The backboard is typically fire rated plywood of some specified dimensions and thickness. Another requirement for some types of circuits is electricity, so be sure to find out the number and type of electrical outlets required. This should be part of the planning for which the equipment requirements are considered as well. If you are ordering circuits to go into an existing equipment room, you may not have to worry about any of this, but check anyway to be sure.

▷ Ask about the availability of "facilities" to deliver the circuit you are ordering. For every single circuit you order, the telephone company must provide a physical place for it to connect in their central office (either a "port" on their switching equipment or a point of termination at a cabling distribution frame.) They must also have available circuit numbers (such as a 7 digit telephone number) to assign to each of your circuits. (Note: If you want an easy number, you need to request it and will pay a premium monthly charge if you get one.) In addition, there must be sufficient capacity on the telephone company's infrastructure (often underground) to physically deliver the circuit into your building and up to your floor. Each circuit requires at least one pair of copper wires (some require 2 pair.) Don't take for granted that all of this is just sitting there waiting for your order. If a new multi-pair cable needs to be run into your building or up to your floor, this can delay the due date by weeks or months. Note: The telephone company may not be able to easily provide you with this information as facilities may not be assigned until a short time before the due date. If your is a "large job" with many circuits ordered, there will definitely be advance checking for availability of facilities.

▷ Circuits are delivered into a telecommunications equipment room or area within your space. Be sure the person you're ordering from specifies this on the order. Otherwise the circuit may be left in a building hallway telephone closet. The telephone companies bring the circuit into your room and leave it there. Be sure it is clearly tagged and labeled. Also write down where it is in case the tag or label disappears. If the equipment to which you want to connect this circuit is also in the room, this may be sufficient. If your equipment is out at a desktop, the circuit must be connected to a cable that runs out to that desktop. There may already be cable in place or it may need to be run. This is done by the company who runs the cable for all telephones and computers within your office. Have the circuit number labeled at the desktop as well.

▶ Plan to be around, or have someone else available, on the day the circuit is to be installed. Remain accessible to the installation person for the entire time they are on site and keep the lines of communication open. Try to be friendly and helpful. Don't act like you know more than the installer does. Field installers regularly get a lot of flack and can be sensitive. Most of them know what they are doing. If you ask for help, you can usually get it. Buy the installer a cup of coffee and strike up some camaraderie. Bringing in a new circuit may require that person to be in the street outside your building, in the basement, in the telephone closet on your floor, in your own telecommunications equipment room or out in the truck. All areas must be accessible or the installation process cannot be completed. If you are not available, the installer may leave and report back to the office a "Customer Not Ready". This means you'll go back on the waiting list for another due date. Not a good thing!

▶ Some types of circuits do not completely work until they are connected to the equipment on which you plan to use them. Ask your equipment provider if this is the case. If it is, coordinate with them so that they can be there to connect and test the equipment when the circuit is installed. If this is not practical, get a telephone number from the person installing the circuit and arrange a time when your equipment person can call to test or "turn up" the circuits.

▶ One more thing, check the bills that come in to be sure that the installation price and monthly charges agree with what you were told when you placed the order. Good luck!

What You May Be Ordering

(Note: The term line and circuit both mean the same thing as they are used in this article. A trunk is a type of line or circuit.)

Plain Old Telephone Service (POTS)/ Dial Tone Line/ Local Line

This is a dial tone line like the one you probably have in your home. These may be brought into a business location to bypass the telephone system for a fax or a modem or just as a bypass telephone. (Note: Faxes and Modems can also use extensions from the telephone system and do not need separate lines.) The POTS line is also called an analog line. This does not mean you cannot send a digital transmission on it, such as from a fax or computer modem. The modem converts the digital signals to sounds – "high tones and low tones" representing the digital signal of 1 or 0.

Information required to order this line includes: directory listing, billing name and address (if you have no previous credit history ordering telephone lines a deposit may be requested and held for several years), installation address, touch-tone or rotary, type of jack in which you want the line terminated (ask the equipment provider), point of termination (exactly where within your premise the line is to be delivered.) and the long distance carrier to be accessed from the line.

Combination Trunks

This type of line is also called a "both way trunk" since it can be used for both incoming and outgoing calls. You order this type of line for a telephone system, usually a PBX or key system. All of the information for ordering "plain old telephone service" above is required. In addition, for a PBX the trunk is usually ordered "ground start" and for a smaller key telephone system it may be ordered "loop start" (see Newton's Telecom Dictionary for more info on what this means.) Ask your equipment vendor whether to order loop start **or** ground start or, better yet, ask your equipment vendor to place the order for you (some will, for a fee, and some won't).

● Direct Inward Dial Trunks and DID Numbers
(Same Type of Service Called DNIS By
Long Distance Companies)

A Direct Inward Dial trunk (also called a DID) is an outside line specially designed for handling and distributing incoming calls to specific telephones within an organization. Again, you will provide all information required for POTS (plain old telephone service). DID trunks are usually "wink start" in the U.S. and may be "immediate start" in Europe. (Again, ask your equipment vendor). A group of DID trunks, lets say 20 trunks for example, is ordered with a group of separate telephone numbers, let's say 100 separate seven digit dialable telephone numbers for every 20 trunks. Now whenever anyone dials your specific telephone number, the telephone company delivers the call to your PBX and then resends the last 3 or 4 digits of the dialed number. This enables your PBX to direct the call to your specific extension. When ordering you also need to specify whether you want 3 or 4 digit "outpulsing" (ask your equipment vendor). Unlike with a POTs line, you are not entitled to a free directory listing with every single DID telephone number, although you may order them to be listed for an additional monthly charge. DID numbers are purchased in blocks of 20 or 100 numbers, usually consecutive. If your organization will be expanding, buy many more DID blocks that you anticipate needing. The number blocks are inexpensive and this can save you from having people within your own office having two separate exchanges (first 3 digits of the 7 digit number).

Long distance companies offer a similar type service caller DNIS (dialed number identification service). In this case the "dialed numbers" are separate 800 numbers and usually delivered over a T1 connecting you to the long distance company.

Centrex

Centrex service may go by a different name depending upon your area of the U.S. Basically, it provides you with the same capability as DID in that everyone in your organization has a separate dialable telephone number. The advantage of Centrex is that you do not need a PBX on site. Instead, you are using the switching capability back at the local telephone company central office. A separate Centrex line is brought in to your premise for each telephone. Again, when ordering new service, all information required on a POTS line order will be needed to order Centrex.

Centrex lines are typically ordered in a group, although you can add more Centrex lines to an existing service. A Centrex line can be similar to an analog POTS line and may be used for the same purposes. It has the added capability of being able to transfer a call to another Centrex line and may have other functions built in such as conferencing capability and speed dialing. Other types of Centrex lines are delivered to work with a digital type of telephone that goes with the Centrex system and will not work with devices requiring analog lines.

T1

A T1 is a high capacity circuit (1.544 Mbps) enabling 24 separate channels or the equivalent of 24 voice telephone lines over two pairs of copper wires. A T1 can be used to deliver combination trunks and DID from your telephone company. It can be used to deliver incoming and outgoing service directly connecting you to your long distance carrier. A T1 can also be used to connect two points within your organization for the purpose of voice communications (sometimes called tie lines) or a combination of voice and data transmission. A T1 may work with two different types of signaling which needs to be specified when you order it (once again, ask your equipment provider). ESF (Extended Super Frame) signaling is used when the application is for voice communications only. B8ZS clear channel signaling may be required when the application is more complex and may involve data transmission.

If the T1 is coming into your PBX you either need a T1 circuit board or, if your PBX cannot take T1, you will order a device called a Channel Bank. In either case, the transmissions being sent over the T1 are being broken out and reassembled into voice conversations or data transmissions by this terminating equipment. The T1 circuit board or channel bank is also called a multiplexer.

You may also need a CSU/DSU (Channel Service Unit/Data Service Unit), yet another piece of hardware to make the T1 work. This hardware may be provided by the local or long distance company or may be purchased from an equipment provider. The advantage of renting it from the circuit provider is that they will take the responsibility if there is a problem with the circuit or the equipment, but this will cost more over time than buying your own equipment.

BRI

A BRI (basic rate interface) line is a type of ISDN (Integrated Services Digital Network) line which may be ordered for a residence or a business. It is delivered on the same pair of copper wires used for a POTS line, but has more than twice the capacity. If you are working from home, but want a business listing in the directory, you pay for a business line (higher rates). The ISDN line (also called 2B+D) has two separate B channels (64 kpbs each) which are the equivalent of two separate telephone lines and numbers. There are actually two 7 digit dialable telephone numbers assigned. There will also be Service Profile Identifier (SPID) numbers assigned to your ISDN line. The D channel enables signaling which makes this ISDN BRI line work in a variety of ways. When ordering it, it is important to specify how you plan to use it which may be for (1) Voice and data, so that you can have a voice conversation going on one line with a data transmission on the other- for example, you are speaking to someone while you are both looking at the same document on your computer screens, or (2) Data only, in which case you will likely be using the capacity of both B channels together to enable faster computer to computer transmission, such as accessing the Internet, the most common use for a BRI line currently. The BRI line also needs terminating equipment, sometimes called an NT-1 adapter. As with other types of circuits, it is always a good idea to ask the equipment provider exactly how to order the circuit and provide them with a copy of your order to be sure you got it right. Another thing to know about a BRI is that, if you are using the line for a data transmission, you are charged an additional cost per minute.

A BRI line must be 3.4 miles or less from the telephone company central office in order to work.

56 kbps

This is a line dedicated to the transmission of data and can be up to 5 miles from the telephone company central office. If you want a dedicated (always connected) circuit for Internet access, you may order a 56 kbps circuit to your Internet service provider.

PRI

A PRI (Primary Rate Interface) is the ISDN equivalent of a T1 circuit. It has 23B channels +1 D channel. You can have up to 19 PRI circuits controlled by a single D channel and do not need a separate D channel on every circuit. This is known as Multiple Facilities Signaling Control. The PRI type of T1 does not require that you designate channels for either incoming or outgoing calls, so you need overall fewer paths than you would on a regular T1 if you are using it for incoming and outgoing telephone calls to your PBX.

Your PBX will need a circuit board specifically for PRI circuits which is different than a T1 circuit board. The PRI also facilitates the delivery of the calling telephone number (Automatic Number Identification) and enables you (with the right terminating equipment) to determine how many rings before callers are answered and how many channels of the PRI are in use at any given time (very useful information for maintaining good service levels for your callers.) As with a T1, the PRI channels can be used for voice and data, the PRI offering more flexibility than the T1. If you are going to use it for data and voice, it is better to bring the PRI into terminating equipment such as a server or router first and not directly into the PBX. PBXs are historically "inelegant" when it comes to stripping out channels of a T1 or PRI for data communications use. The typical PRI, like the T1 uses what is called B8ZS clear channel signaling.

● ADSL and SDSL

This stands for Asymmetrical or Symmetrical Digital Subscriber Line. They enable very high transmission capacities (higher than BRI ISDN, almost as much as T1, but at a much lower cost) over the copper wires currently in place. DSL will be used for video transmission as well as Internet access and voice communications.

■ "Who You Gonna Call?"

Every day there are more and more companies from whom you can buy these services, but for the sake of simplicity, let's identify two separate types of companies whose service offerings are beginning to overlap: Local Telephone Companies and Long Distance Telephone Companies.

When getting ready to place an order, the first step is to find out if your organization already has a relationship with someone at one of these companies. If so, that will be the person to call. It is also good to find out if you may have a contract with one of these companies to be sure that the services you are ordering benefit from contract pricing. This covers not only installation and the monthly charge for the line, but usage charges as well. For example, if you order a POTS line, make sure that you tell your long distance carrier so that calls made from that line will be billed on a cost per minute basis at your contracted rate.

If you have no established contacts, the best way to get the correct number for the local telephone company is to dial directory assistance (411 in most areas) and ask for the telephone number of the local business office for "Name of the Telephone Company" handling business (or residence) orders. This is separate from the billing department. Then call the number and start the process. Don't get frustrated if you don't get the right person on the first try. The service offerings are complex and different groups within the telephone company are trained on different types of circuits. By the local telephone company, we're referring to the one who has been around the longest, such as Verizon, Bell South, etc. In most areas now these companies have competition, so you can shop around. The competitors may be set up with a somewhat different structure than that of the original Bell Operating Companies we've mentioned, but you should still be able to get a telephone number from directory assistance.

Start the same way for circuits from long distance service companies. Get their telephone number from directory assistance and call them. Again, it is much better if you already have some contact. You call a long distance company for long distance service or for a circuit connecting two points in two different geographic areas not served by the same local telephone company.

Sometimes you have a choice. For example, a T1 circuit from New York City to New Jersey can be rented from Verizon local telephone company or from a long distance telephone company. There are also competitors to Verizon called CLECs (competitive local exchange carriers) from whom you can rent the same types of services.

Both local and long distance service (also Internet access and cellular service) are also sold by Authorized Resellers and Agents who can sell the services of one or more local or long distance carriers.

In some areas, you may also buy some of these services from the Cable TV company.

Negotiating Next Year's Long Distance Contract

Not for people with previous history of heart problems!

If you haven't renegotiated your long distance contract in a year or more, don't hesitate! The rates may have come down. Even if your old contract has not yet expired you may be able to renegotiate new lower rates with your long distance service provider.

Here are some suggestions to help you with the process.

▨ What Are You Spending Now?

The hardest part of the entire exercise may be determining what you're current rates are, particularly if your contract was written several years ago when the rate structures were more complex. If you don't find out what you're paying now for each type of call, it will be hard to validate how much you'll be saving with a new contract.

This is not something you can rely on the long distance company to demonstrate once the new rates are in effect. They should be able to, but it never really works out that way, so plan to do it yourself. We prefer basic mathematical computations to validate savings projections.

Current total minutes in each category of call multiplied by *current* cost per minute minus current total minutes in each category multiplied by *new* cost per minute should give a good approximation of projected savings. If the volume goes up, savings should increase proportionately.

Determine the cost per minute, initial time period and subsequent time billing increments for the following types of calls:

▶ Intra-LATA (nearby toll calls)

▶ Intra-state

▶ Interstate

▶ International (get information specific to each country you call)

The above types of calls may be billed differently depending upon whether the call is:

▶ Inbound (toll free) or Outbound

▶ Made on a dedicated circuit from your PBX (like a T1) or switched outside lines (lines from your local telephone service provider)

▶ On net (from one of your locations to another if both have dedicated T1s to the same long distance company) or Off net.

You may also see day, evening and night rates or different rates depending upon the distance from your office that you're calling. This type of billing has largely disappeared, so don't get too focused on figuring out how many minutes of calls you have in these categories (unless your operation is open at night or on weekends). Just lump the total minutes together in the categories above and do the math to find an average cost per minute.

Also find out what you're paying now for:

▶ Directory Assistance (cost per call rather than cost per minute)

▶ Calling Card (cost per minute plus a per call surcharge)

We suggest that you obtain current rates from the billing person at your long distance company rather than from the sales representative. In fairness to the sales reps, rates are so complex and no two customers are billed in exactly the same way, so it is often not straightforward to figure out.

The rate shown next to the detail of a call on your paper bill is seldom the real rate, since discounts that show up in various forms and formats often apply. Sometimes the discounts are not even shown on the bill but sent to you as a separate credit that you must send in with your next bill. The discounts are frequently different for each category of call.

If your contract pricing is based upon a "discount off of a tariff" ask to see a copy of the tariff to validate the pre-discount rates.

Verify That You Were Billed Correctly In The Contract That Is Ending

While you're negotiating your new contract, it's a good idea to validate that you were billed correctly under the current contract. You can try doing this yourself (keep TUMS handy), ask your long distance company to demonstrate it mathematically or hire an independent Telephone Bill Auditor with a specialty in Long Distance Billing to do an audit for you. The auditor is usually paid a percentage of a refund he gets back for you.

Many people are of the impression that when you sign a contract for a certain rate that the rate applies for the life of the contract. This has not historically been the case for long distance contracts, which makes it even harder to verify that you were billed correctly since there may have been several different rates in effect for each type of call over the time period of the contract.

Fix Your Costs And Ask For A "Rerate" If You've Been Paying Higher Than Market Rates

Insist that with a new contract the rates in all categories shown above will be fixed until you renew. You'll be surprised how resistant some of the major long distance companies are to doing this. It is also interesting to note that while you have been continuing to pay above market rates, your long distance company has been selling the same service to new customers for much less than you're paying. If you know to ask for it, they may be willing to "rerate" your calls for some period of time with the new lower rate you negotiate. In effect, the new contract you negotiate is deemed to have been effective at some date in the past. This can often generate a significant credit on your bill.

Create Competition

While we are all for loyalty in business relationships, your present long distance service provider cannot offer you their very lowest rates unless you are considering going to the competition, so make sure you get competitive bids.

We suggest creating a Request for Proposal as you would for any other major purchase, enabling you to ask detailed questions of each company and to compare responses.

Negotiate All Of The Following As A Part Of Your New Contract

▶ *Calling Card Surcharges* – they are negotiable and in some cases may be waived although the cost per minute will then be higher.

▶ *Lower Billing Increments* – Some long distance companies provide one second billing increments. One of the major long distance companies has 18 second billing increments. That means for every call you pay for 18 seconds whether you use it or not. Don't assume that all calls within your contract have the same billing increments. Get this in writing for each type of call, as part of your contract.

▶ *Installation Charges* – If you are changing long distance companies or need a dedicated circuit installed (like a T1) ask that the installation charges be waived. You may also ask for a credit to offset the cost of any hardware you must buy for the T1 such as a circuit board for your PBX.

▶ *Minimum Usage Commitments* – If the rates you negotiate are based upon minimum usage commitments, make sure that you can easily meet them. Some long distance companies quote minimum usage pre-discount and some post- discount which can be confusing, so find out which is the case. Also ask for a "business downturn clause" that will enable you to keep the same rates if your usage drops due to changes in your business. It may also lower your commitment level.

▶ *Ability To Renegotiate Prices Within The Life Of The Contract* – As we mentioned above, the long distance companies often increase contract prices during a contract. So turn the tables on them and ask for the ability to request lower prices during the contract if you find out that the rates for the overall marketplace have gone down.

▶ *Toll Fraud Prevention* – Find out what your liability is should someone hack into your telephone system to make fraudulent long distance calls on your lines or make unauthorized use of your calling cards. Most carriers have different levels of "insurance" for a nominal cost, so make sure you cover this as the threat is real.

▶ *Options for Billing Formats and Management Reports* – Validation and administration of long distance bills can be nightmarish. Look for clear simple billing formats that enable you to easily see the cost per minute on each category of call (if you believe you are actually looking at such a document, pinch yourself to make sure the experience is real!). Find out if you can get access to billing information during the month from a web site. (Note: Some independent companies such as Comware Systems, Inc. 203-326-5500 can also provide web based access to your billing information they receive directly from the long distance company. They can also send e-mails to different departments within your organization to let them know how much they're spending on long distance.)

▶ *Customer Service Team* – Find out who will support you with billing questions and service problems. Don't rely on just one person as account reps for long distance companies often leave for greener pastures or are reassigned to other accounts.

▶ *Commitments For Certain Types Of Calls* – Try to stay away from committing a certain percentage of your calls be domestic and another percentage international. Some long distance companies ask for this, but it can be hard to predict and manage.

▪ It's Not Just Long Distance Anymore

While long distance contracts are still typically negotiated independent of other services, the trend is for one company to offer some or all of the following services. The more of these you buy from one company, the lower your rates may be. Since the ability to offer these different services largely resulted from the relatively recent merger with or acquisition of other companies, you may notice that it still seems like you are dealing with separate companies, including separate billing systems and different account representatives.

These additional services include:

▶ Local Telephone Service (outside lines and calls)

▶ Cellular Telephones

▶ Pagers

▶ Data Communications Circuits

▶ Frame Relay Services

▶ Videoconferencing Services (often use what are called SDS or switched digital service rates)

Happy Shopping! Don't weaken! Negotiating long distance contracts is not for the faint of heart.

Purchasing Calling Card Services

Beware of the Bong!

Calling Card costs are often overlooked in the negotiation of a contract with your organization's long distance service provider. The savings realized from that new low cost-per-minute for calls placed from your office can quickly evaporate if Calling Card costs have not been agreed upon.

The Calling "Card" may be an actual plastic wallet-sized card or simply a number. In either case a PIN (personal identification number) is used along with the Calling Card to validate that the person using the card is authorized.

The Calling Card may be used at most public telephones in the United States (and some outside the U.S.). It may be also be used when placing calls from a business or residential telephone, if you do not want your call billed to the person or organization whose telephone you are using.

■ Calling Card Costs

● Cost-Per-Minute

One component of the cost of using a Calling Card is the cost-per-minute (also called the "rate"). If you have negotiated a favorable cost-per-minute for long distance calls dialed directly from your business location, the cost-per-minute for Calling Cards should match this rate if you buy from the same provider. If your organization is large enough to warrant both dedicated (directly connected through a high capacity circuit) and switched (accessed via your local telephone company lines) long distance service, it is likely that your Calling Card rate will be the same as the switched rate, rather than the lower dedicated rates.

As with other dialed calls, calls using Calling Cards have different rates for intrastate, interstate and international calls to different countries.

Rates for local and intra-LATA (nearby calls outside your local area) should also be included in your agreement.

As with all long distance contracts, attempt to negotiate rates that are fixed for the duration of your contract. If your rates are based upon a percentage discount from a "tariff rate" and the tariff goes up, your rates will correspondingly increase. While it appears that tariffs will be going away soon, for the moment, they still exist, as do the "discount off tariff" billing arrangements.

Billing Increments

Calls are charged for an initial period, such as 18 seconds, and then in increments, typically of 1 second or 6 seconds. Higher initial periods and increments add to your costs since you may pay for seconds you have not used.

Long distance providers may assign a longer initial period to a Calling Card call than to other calls, up to a full minute! This means that if you make a call and the distant end answers, if you speak for fifteen seconds, you pay for 60 seconds.

Negotiate for a low initial period and the smallest possible billing increments.

Surcharges

Each Calling Card call carries what is called a "surcharge", which makes it sound like something that is mandated by the government, but is actually just an addition to the cost of every call that is negotiable. We've seen surcharges of up to $2.50 per call (even though the cost of the actual call may be much less) down to about 45 cents per call.

It is also possible to negotiate Calling Card rates without a surcharge, but you will pay a higher cost per minute. This still may be a better arrangement overall, depending upon the average length of your calls.

The surcharge is sometimes referred to as the "bong" charge, after the sound you hear when placing a call before the automated system asks you to enter your Calling Card number.

Pay Telephone Charges

While strictly speaking, pay telephone charges are not considered part of Calling Card rates, they are nevertheless charges that are added to your bill when you use a Calling Card at a public telephone.

The justification for these charges is that your long distance service provider must pay something to the company providing the pay telephone. While it makes sense that the pay telephone provider should not provide anything without charging, what the costs will be is something that should be part of your Calling Card contract negotiation.

Who Sells Calling Cards

Each of the "big three" (also called Tier 1) long distance service providers sells Calling Cards. AT&T, MCI Worldcom and Sprint (Foncards). Other long distance providers (called the Tier 2 providers) typically have their own Calling Cards as well.

While most long distance companies are also selling local telephone service, the traditional local telephone company also provides Calling Card services. (This company may be known as the ILEC – incumbent local exchange carrier.) If your organization makes a lot of local calls or long distance calls to nearby areas, it may cost less to make these calls using a Calling Card from your local telephone company.

We do not recommend issuing two separate Calling Cards to a staff member with instructions to use one or the other, depending on the destination of the call. This creates confusion that quickly offsets the benefits of any cost reduction opportunities.

Calling Cards are sometimes sold as part of a promotion with a Credit Card company, often with apparent added benefits such as Airline Miles. Look at these offers closely as they bear hidden costs that offset the benefits.

■ Unsolicited Cards

It is not uncommon for unsolicited Calling Cards to arrive in the mail, often from your traditional local telephone company. These may inadvertently get into the hands of staff members who use this card and incur significantly higher rates than if they used the Calling Card that is linked to your organization's overall long distance provider agreement.

Review your bills carefully for evidence of use of unsolicited Calling Cards.

■ Ease-Of-Use

In most cases, it makes economic sense to purchase your Calling Cards for long distance calls from your long distance service provider.

However, it is important to be aware that there are some variations in the ease with which Calling Cards are used. This is something to be particularly sensitive to if you are changing from one long distance company to another.

There are some instances where Calling Cards from AT&T are easier to use, such as from an AT&T provided pay telephone or AT&T telephone in an airport club such as the Admiral's Club. In some cases fewer digits need to be dialed or the actual Calling Card (if you carry it with you) can be "swiped," further cutting down on the digits dialed.

Most Calling Cards require the user to dial a toll free number (11 digits) plus the called telephone number (another 11 digits for a call within the U.S.) and then a PIN (personal identification number) (14 digits) for a grand total of 36 digits – not really too convenient by any stretch of the imagination. Some long distance providers make it easier if you are calling back to your home office or enable you to preprogram "speed dial " for frequently called numbers.

Most service providers make you start all over again if you make a mistake!

Geographic Coverage

If your staff travels a lot, it is important to be sure that the Calling Cards will work in other countries around the world. At this point, AT&T has the widest coverage, with over 200 countries. MCI Worldcom and Sprint each cover over 80 countries. This may be sufficient depending upon where your travels take you. The Tier 2 companies' Calling Cards may or may not cover where you're going, so check before you leave.

Get explicit instructions of how your staff will use their Calling Cards from hotel and pay telephones outside of the United States, particularly those that may not yet have touchtone service. This can be written into the agreement with your service provider.

Options For Billing

Details of Calling Card calls are typically broken down by Calling Card number and provided on the Billing Detail. This may be on paper or on a CD-ROM, depending upon how you get your bill detail. A few long distance providers advertise call detail information available on a web site.

Some organizations prefer to have an individual's Calling Card use billed to a corporate credit card account that the staff member uses for other business expenses. If this is something you'd like to do, be sure that the service provider can arrange it. Even if call details are billed to individual corporate credit card, be sure that a summary report will be available for management purposes.

Reporting Options

When making decisions about purchasing Calling Card services, it is important to have good information about what your organization's Calling Card usage actually entails.

Request monthly reports not only by user but total minutes for different types of calls such as local, intra-LATA, intrastate (for each state), interstate and international (U.S. to each other country/ each other country to the U.S./ and from one country to another outside the U.S.). Long distance service providers are reluctant to provide fixed rates on all international Calling Card calls. If you can provide your most frequently called ten or twenty countries and the expected volume of calls to each, there is a greater likelihood that you will be able to negotiate a fixed rate which will not change for the duration of your long distance contract.

Keeping Track Of Users

It's your responsibility to keep track of who is using each Calling Card billed to your organization's account. This is best checked on a monthly basis.

Have a procedure in place to invalidate the Calling Card number when the user leaves your organization. If you don't and the person continues to use the card, you're responsible.

Many large organizations keep spare Calling Cards for new employees to use, so checking to be sure these spare cards have no usage is another part of this administrative task.

Toll Fraud Protection

Unfortunately fraudulent use of Calling Cards is a big business. Someone in an airport can overhear you if you speak your card number (sometimes needed if an operator intervenes). Other Calling Card thieves look over your shoulder with binoculars to see your card number or can translate the touchtone signals they hear you dialing.

To limit your liability if this happens, sign up for one of your long distance service provider's toll fraud insurance plans.

Also find out what the procedure is for quickly replacing a Calling Card if the card itself is actually stolen or lost.

Advanced Capabilities

Ask your long distance service provider to explain their network's advanced capabilities to you in terms of services that may improve the way your organization using Calling Cards. For example, can your Calling Card users access a directory of frequently dialed numbers that is maintained in the service provider's network? Find out the costs, including the cost of administration and try to experiment with anything new before rolling it out on a large scale.

Purchasing a Maintenance Contract for Your Telephone System

What do you mean you'll have someone here next Thursday?

Most organizations give considerable thought and energy to the purchase of a new business telephone system. The Purchase Contract is carefully negotiated and often revised. Unfortunately the accompanying Maintenance Contract on the system is frequently ignored and viewed as the less exciting part of the process. When problems crop up down the road, you'll be glad that you invested some time on the maintenance issues to keep your system running smoothly and to obtain a quick response when it's not.

■ Definition

The Maintenance Contract on a business telephone system ensures that when there is a problem with the system it will be quickly and adequately diagnosed and fixed. The Maintenance Contract may also be called a Maintenance Agreement, Service Contract or Repair Service Agreement. It describes all of the maintenance services to be provided along with the terms and conditions. It covers both labor and replacement parts for repairs.

Unless your system is large enough (1,500+ people) to warrant an on-site technician, if a repair problem arises, you call and report the problem to your Telephone Installation and Maintenance Company.

A repair problem ranges from a single telephone not working, to an outside line not working to the entire system crashing, in which case no incoming or outgoing calls can go through.

Don't confuse repairs with system changes. Things such as rearranging telephones or changing the extensions that appear on them are usually not included in a Maintenance Agreement, although in some cases, this may be offered as an option for an additional fixed monthly charge.

■ Who Maintains Your Telephone System?

The company from whom you purchased your telephone system is almost always the company who installs the system and provides maintenance, particularly when the system is new. We're calling it the Telephone Installation and Maintenance Company, but you may know them as your PBX supplier, PBX vendor or Interconnect Company. (PBX stands for Private Branch Exchange.)

The company that installs your system is most familiar with your premises and your particular system (no two are set up in exactly the same way!). It makes sense to have them provide maintenance support as long as they are doing a good job at the right price. The purchase price of a new business telephone system usually includes a Maintenance Contract for one year or more.

If your Maintenance Contract is up for renewal, you may want to diplomatically obtain bids from other companies to keep your current maintenance company on its toes. This can be done informally, but if you wish to accurately compare pricing and services, a Request for Proposal can be prepared. Seeing the Request for Proposal tends to "sharpen the pencil" of the companies that are responding, often resulting in a reduction in the cost. Your alternatives depend upon the manufacturer of your system and how many Telephone Installation and Maintenance companies support this system in your local area. This makes a case for purchasing a system widely in use, since this increases the likelihood for future support alternatives. Quality of support from Telecommunications Installation and Maintenance Companies who claim a national presence can vary dramatically from one part of the country to another, so always check them out locally, calling a number of references.

To determine the alternatives for Maintenance Support for your system in your area, contact the manufacturer (all have websites), but also check the Yellow Pages (under the manufacturers name or look at the large ads) and ask your business associates. The manufacturers often play favorites or just don't know and may neglect to mention all possible alternatives.

Some manufacturers have Telephone Installation and Maintenance companies as part of their organization. Again this varies from one area to another. In many cases, the manufacturer-owned entity competes with other companies authorized to sell and support the same system.

Pricing

While the price for the Maintenance for the first year or more is buried in the system purchase price, it is important to determine how the Maintenance Contract will be priced when it is up for renewal. (You can ask for the cost of maintenance to be detailed in your purchase contract, but this won't necessarily lead to an overall better price.)

You can negotiate the renewal price when you purchase the system. If you agree to stick with the same maintenance company, you may be able to get a lower price, but limit your flexibility in the future which can have a far greater cost. We suggest not locking yourself in until you have some history with the maintenance company. Also remember that companies change, often dramatically, as they merge or are spun-off or as key employees leave. If the technician most familiar with your system moves to another company, you may be better off moving with him than having a new technician who is unfamiliar with your organization and your system.

In most cases the price for the Maintenance Contract is computed based upon the size and complexity of the telephone system. As other items are included, the price will be affected accordingly, as it should be. A reality in the current marketplace is that Telephone Installation and Maintenance companies are making very slim profits on the sale and installation of systems and view the Maintenance Contracts as an important profit center.

The most common method of arriving at a Maintenance Cost is to multiply the number of active ports in the system by an amount such as $5.00 per month. Each telephone in the system uses one port and each outside line in a system may use one port. Active means that the ports are actually connected to a telephone or an outside line or are sitting in readiness to be connected (spare ports). So a system with 200 telephones and 25 outside lines would cost $1,125. monthly for the Maintenance Contract (225 ports x $5.00 per port per month). This is an oversimplification, since most telephone systems today are more complex with high capacity circuits connecting to the outside world and links to other systems such as Voice Mail, Interactive Voice Response and other computer applications. These add to the cost. Nevertheless, there is some formula, and knowing how the maintenance cost is computed is important both for negotiating and for keeping the cost accurate in the future.

What happens when you add ten new telephones and the associated circuit board? How do you keep track if they have one year of maintenance that does not expire at the same time as your other telephones and circuit boards? How can you simplify things so that all telephone system components will have maintenance up for renewal at the same time without costing you more? These questions have no easy answers, but the cost of not figuring this out can be significant.

Some organizations reduce the expense of a Maintenance Contract by omitting the actual telephone instruments from the Contract, keeping some spare telephones on hand for replacing those that malfunction. They also may send the broken ones in for repair. While this takes more work on the part of in-house staff, it should be looked at to see how much it can lower the overall expense of the Maintenance Contract. Try to factor in the cost of the time for your staff person doing this.

Another element of the cost arises when your Maintenance Contract comes up for renewal, even if you plan to stay with the current service company. Someone needs to check to be sure the pricing is still based upon an accurate count of the active ports. Suppose your organization has gotten smaller? You may still have ports in the system no longer connected to telephones and may elect to have them deactivated to lower the maintenance cost. (In order to do this, there may be a cost to reassign the telephones to different circuit boards, enabling the service company to deactivate entire circuit boards.)

What Is Covered?

A Maintenance Contract usually covers both labor time and the replacement of any system component. It is important to have the quantity and description of each system component listed (in English, as well as in part numbers or unfamiliar acronyms.) If related systems such as Voice Mail or Call Accounting are covered by the agreement, both the systems and their components need to be described. As mentioned above, you may wish to explore the option of leaving the telephone instruments out of the contract, keeping replacements on hand instead. Also find out if cable is included as part of the Maintenance Agreement and if not what the cost will be if it is cut or proves faulty.

Hours Of Coverage

For most organizations, maintenance coverage during business hours is sufficient, but if you operate round the clock you may want to opt for a 24 hour a day, 7 day a week contract. You can also negotiate a business hours only contract and pay for labor time outside of normal business hours if you need it. Just be sure that the replacement of all equipment is covered regardless of when the failure occurs.

Response Time

It is reasonable to expect that when your telephone system fails completely (this should not happen very often!) that the response will be almost immediate. This makes a strong case for having a Maintenance Company who is located nearby. If a major part of the system fails the response should also be rapid. Most Maintenance Companies are reluctant to commit to shorter than a two hour response time in the event of a major failure, but practically speaking, it is likely that a good company will show up faster than that.

For minor problems such as a single telephone being out of service, a response time of one business day is reasonable.

While telephone system manufacturers and Maintenance Companies advertise that their systems can be remotely diagnosed and sometimes repaired, we find that this rarely happens and there is seldom a substitute for a site visit.

▪ Remote Diagnostics

Remote diagnostics may be able to detect a problem and anticipated the need to send someone out to fix it, often before the problem is reported. Dialing into a telephone system regularly and checking for trouble requires both time and expertise, so if you're anticipating this from your service company it is important to be clear on the expectations, the process, the schedule and the costs.

▪ Failure Definitions

Sit down with your Maintenance Company to define minor and major failures. Some companies consider it a major failure when the President's telephone does not work.

▪ Software Upgrades

We suggest that you ask your Service Company to include the cost of keeping your system software current as a part of the Maintenance Contract. While this will add to the cost, it will be less painful than being surprised with a separate software upgrade cost for many thousands of dollars at some point in the future. Keeping the software current ensures that you will benefit from any improvements and that your system will remain easily supportable. Technicians may come and go and the new ones are typically trained on the latest software version.

Traffic Studies

A traffic study checks the volume of calls on each of your outside lines (or high capacity circuits such as a T1) to determine whether you have a sufficient amount of lines (or too many). It's a good idea to have your Maintenance Company do this once a year. We suggest adding it into the Maintenance Agreement.

Liaison With Telecommunications Service Providers

Who will call the local or long distance telecommunications service provider if there is a repair problem with one of your outside lines? If this is not included as part of the Maintenance Contract, you may find your staff spending many frustrating hours on-hold trying to get an outside line fixed. We strongly suggest that this be included as part of the Maintenance Contract, despite an additional cost.

Repair Log

One of the ways to compare Maintenance Companies is by their process for maintaining a repair log. The good ones will keep a good record of all problems reported, what the outcome was and what caused the problem. This can be valuable in terms of negotiating your Maintenance Contract next time around. If you don't have many service calls, you may be able to negotiate a lower cost than someone who frequently calls for service.

Time And Materials

While you can obtain maintenance for your telephone system on a time and materials basis, we suggest having a

Maintenance Contract instead as it gets you higher priority with your Maintenance Company and can pay for itself if some major system component fails and needs replacement.

The Request for Proposal Approach to Purchasing a PBX

You can never ask too many questions.

The intent of a Request for Proposal (RFP) document is to assist an organization in making an intelligent purchase and to provide companies competing for the business with a common description of the requirements.

One objective is to solicit proposals that can be easily compared in terms of price and other characteristics of the product being purchased. A second objective is to be sure that what is purchased is complete and that there are no last minute surprises.

A business telephone system is a complex purchase. While we are in favor of and regularly use the RFP approach to support the purchase of a PBX, it is important to recognize that a PBX is not a commodity in that no two systems from different manufacturers can be purchased in exactly the same manner from a hardware, a software or an operational perspective. Other factors such as the outcome of a system demonstration, feedback from system users and the relationship with the system provider are as important as the RFP response in the selection process.

It is also important to note that a PBX is but one component of a business telephone system. Other elements include cable, network services (outside lines for local and long distance calling), Voice Mail/Voice Processing systems and links to the computer network for a variety of applications. Separate but related RFPs can be prepared for each of the different system elements, since not all are necessarily sold by a single provider.

Get Buy-In From Others In Your Organization

We're starting at the point where a case has been made to management that a new PBX must be purchased and that a budget is in place.

Many organizations purchase a PBX without a clear picture of what is to be accomplished and often wind up with a new system that works no better than the old one. A new PBX is typically purchased when an organization moves or when it outgrows the existing system. Less frequently the PBX is replaced to gain some operational improvements it is expected that new capabilities will deliver.

We suggest getting representatives from your organization's various departments to participate in putting together the requirements.

Discuss and document what works about your present telephone system, what functions you can't live without and how the handling of incoming, outgoing and internal calls can be improved. What are the issues from the perspective of the callers, the employees, the system administrator and upper management? Different people within your organization will likely have different views of the objectives and how to measure whether they are met. Draw diagrams showing the path a telephone caller takes into your company and what happens when the call gets there. If you're not clear what is happening currently, this is the time to find out!

While no PBX meets all criteria, the more participation you have during this initial stage, the greater the likelihood of a successful outcome.

Putting Together The RFP

The remainder of this tutorial will provide you with an outline of sections to be included and a few sample questions suggested for the PBX Request for Proposal.

Introduction to Your Organization and The Project

Provide a clear and interesting description of your organization. What is your business? What variables affect the success of your business? How many locations do you have and how many people work at each? What are the key departments and their functions?

Why are you purchasing a new PBX and what does the current project include (for example, are you moving, too)? What are your objectives and your timetable? Does the proposer stand to gain additional business if this project is a success?

Current Environment and System Operation

With what type of telephone system are your currently operating (manufacturer name, model, maintenance company)?

How many telephones do you currently have? What type and how many outside lines are connected to your current system?

How does your telephone system operate now for handling incoming, outgoing and internal calls.

What system functions work well and which one's are in need of improvement? (For example: The current system enables us to conference only three people, but we need the capability to conference up to 5.)

Objectives for the New System

Provide more detail in terms of your expectations for the new system. Does Management want better reporting in terms of how callers are being handled? Is your current system unable to deliver certain functions such as displaying the telephone number and name of the person calling (Caller ID)? Are you replacing PBXs at each of your locations to facilitate an intra-office four digit dialing plan?

System Specification

Here's where you put the number and type of what you want to buy. As we mentioned earlier, since telephone systems are put together in different ways, your first specification will need to be refined before you actually purchase a system.

For example:

▶ *Number and Type of Telephones.* While you can provide the count of telephone instruments, the type may vary from one manufacturer to another since they each make telephones with varying numbers of buttons for different functions. A ten-button telephone from one company may suffice where you'd need to buy a twelve-button from another supplier. Here's where the description of how you plan to cover calls comes into play. If you indicate that all message taking telephones need an appearance of both extensions to cover calls for ten different telephones, this will begin to indicate the size of telephone that may be right for this particular job category. The cost of one telephone in a manufacturer's product line can vary significantly from another. Display telephones cost more than non-display, but are almost always warranted.

▶ *Number of Switchboard Consoles.* Most PBXs require at least one switchboard console. Decide what role the switchboard will play in your overall call coverage plan.

▶ *Number and Type of Outside Lines.* We suggest providing the number and type of outside lines and letting the proposer fill in the number and type of circuit board needed to support them, since the number of outside lines handled by a circuit board can differ from one manufacturer to another.

▶ *Links To Other Systems.* If you need links to other systems such as voice mail, this will take up space and require circuit boards in the PBX.

▶ *Growth Capability.* You can purchase growth in terms of buying circuit boards for telephones and outside lines or just purchase space in the PBX cabinet into which circuit boards can be inserted later. It is important to understand how the system is put together and what is needed to expand it.

▶ *System Configuration.* Understand how the proposed system is physically put together.

Request a diagram of the control cabinet for the proposed system and an explanation of each system component.

▶ *New System Operational Capabilities.* Compile a list of all conceivable telephone system functions (you can get these from most PBX brochures) and make sure you define each since different manufacturers may handle similar sounding functions with very different methods. Newton's Telecom Dictionary (from 1-800-LIBRARY) can help with the definition process. Think about how the system functions come into play for either incoming, outgoing and internal calls. Define the functions in terms of exactly how you expect them to operate. While this is tedious, the more explicit you are, the fewer surprises you will have when the system becomes operational.

Think about different capabilities in terms of Switchboard Functions, Telephone User functions, System Administration and Management reports.

Describe your expectations in terms of how call coverage will be handled in different departments both during and after business hours.

Since most PBXs are now sold with voice mail, a description of the voice mail as it ties in with the PBX is also in order.

■ Other Sections of the PBX Request for Proposal May Include:

▶ Project Timetable

▶ Installation Expectations

▶ Maintenance Expectations

▶ Disaster Recovery Requirements

- Equipment Room and Environmental Requirements

- Training Requirements

- System Documentation and Record Keeping Expectations

- Information About the Company Submitting the Proposal

- Information About the Proposed System

- Terms and Conditions Required by Your Organization for any Large Purchase

Tips For Improving The PBX RFP Process

From the very beginning of the process, think about developing the Request for Proposal so that you will be able to get through the responses in the quickest and most effective way possible.

When you are certain of precisely what you want, whether it is a necessary PBX function or a requirement for labeling jacks, instead of a question, just make a statement that this is a requirement. The respondent can then state that they comply and may provide a longer answer only if they do not.

Of course some areas will require questions where you want to compare systems such as, "how many PRI circuits can your system accept at maximum capacity?" Try to phrase the questions so that the answers can be simple and unambiguous.

Instruct the proposers who want to provide a lot of boilerplate verbiage to support the response to a particular item to reference it in the response, but to put in an appendix at the end. If long responses are included they tend to interrupt the flow of the proposal and make comparing responses from different suppliers difficult.

You may think you know the selection criteria in the beginning, but the real criteria may not be apparent until you start comparing systems. So don't create unrealistic expectations with the proposers in terms of how the selection process will work.

There may be a PBX that does everything you want it to, is priced competitively and is supplied by a company with outstanding support, but if the CEO of your company hates the telephone that will sit on the desk, that company will not be awarded the business.

We suggest going to see demonstrations of systems to be considered very early in the process, right after you have defined your requirements and expectations. Then you can see how the different systems look and can experience using them. There is no point is asking a supplier to spend the time and energy responding to a Request for Proposal if, once their system is demonstrated, they are out of the running because someone does not like the telephone.

People are often surprised to learn that the telephone instrument itself (the part of the PBX most evident to the users) is the deciding factor in most PBX selection processes.

Even if you know exactly what system you want to purchase, the Request for Proposal is still an important part of the process, ensuring that the system will be purchased with all appropriate components and capabilities includes. Once you have the system selected, you may also wish to shop the price with multiple suppliers of the same PBX.

Big Telecom Buys

Where Mistakes Cost More!

You're relocating; you're expanding; you're hoping to improve the handling of your organization's telephone calls. If any of these describe your circumstances, you may now be faced with making the BIG TELECOM BUY!

Whether it's a complete new telephone system or adding a major component such as voice mail, here are some suggestions to get you pointed in the right direction.

Don't Go It Alone

People have high expectations for their telephone systems and are often disappointed. A good way to avoid this is to get as many people as is practical involved in the process of selecting the new system. A small, thoughtfully selected committee will improve the likelihood of success. It avoids having one individual make the decisions for the organization based upon his view of what everyone needs, but do appoint one person as the group leader. Users are more willing to accept inevitable shortcomings if they have participated in the process.

Select representatives from information technology, human resources, sales and marketing, customer service, secretaries for the top executives and the switchboard operator.

Don't Think About Technology First

We often see too much emphasis on the technology behind the systems without any understanding of how well the system will operate in a particular office environment. For example, the appearance and ease of use of the telephone instrument is a critical component of the selection process. This is the only part of the telephone system that everyone will see and use every day! Yet the telephone instrument is not often a focus of the evaluation process.

All manufacturers of business telephone systems are very tuned into the need to keep their product lines technologically current. Each manufacturer is somewhere in the process of accommodating the latest technology (even before they are sure it will be important to buyers). Business telephone systems are changing. Some are the more traditional and time-tested products such as the PBXs from companies such as Avaya (formerly Lucent), NorTel, Siemens, NEC, Mitel, and Ericsson. Others are the newer breed, such as Cisco, often more closely integrating with the organization's computer network.

In either case, make sure that the product you buy is not about to be dropped from the manufacturer's product line.

Inquire about the new technologies and how the particular manufacturer is developing the product to accommodate them.

The technology may have to do with how communications signals are sent and received. VoIP (Voice Over Internet Protocol) is a recent trend that is projected to take off (no guarantees!). The system manufacturer will likely have a circuit board that can accommodate this that must be purchased along with the system.

It may also have to do with how the system interacts with other technology systems in the office. The broad area currently known as Convergence encompasses many separate capabilities such as being able to dial out from your computer based contact management system, to having a screen of information about the caller appear on your computer screen as the call arrives.

▪ Find Out How Things Are Working Now

In wanting to move ahead with a new system, organizations seldom focus on how their present telephone system is working. In not doing so, much valuable information about what works and what doesn't is often lost.

Get into the details! Exactly how do the intercoms work? How is someone on a call notified that another call is waiting? What do you see and what do you hear? How many buttons are on the existing telephones and how are they used? How many buttons are used for speed dialing? How do users retrieve their voice mail and how do they get to someone else in the office after doing so? Don't assume all departments work in the same way, as they seldom do.

What are the limitations of the present system that you hope to overcome with the new system? If you don't identify them and document them, opportunities to improve things may be overlooked.

Evaluate The Importance Of New Capabilities To Your Organization

When evaluating systems, it's important to see how the products compare in terms of how system functions work. Before you can prioritize what's important, make a list of all possible capabilities you may get with a new system that you don't have now. Each item needs a clear non-technical explanation describing what it is and how it works so that all members of your telephone system selection committee will understand.

For example, if you're buying a new voice mail system, do you want to include capability for Unified Messaging so that voice mail messages appear on the user's computer screen along with the e-mail? Who will use this in your organization and how practical will it be to deploy? Should you test it out with a pilot project first (always a good idea.) What kind of cooperation will you need from the Information Technology department?

How about Caller ID, if you're buying a new telephone system? Do the type of outside lines you're using deliver the Caller ID (calling telephone number?) How will you use this information?

Put It In Writing

Once you have decided what functions are important and how you plan to use them, it's important to write it all down. Write in clear and non-technical language, since "technical" or "telephony" terms often mean different things to different manufacturers and suppliers. Incorporate this description into a Request for Proposal that asks questions about installation, maintenance, ongoing support, scheduling and other issues relevant to the purchase. The Request for Proposal will also include a count of how many telephones of each type (recognizing that no two manufacturers put their telephones together in the same way), the number and type of outside lines and expansion requirements (even if you don't plan much growth, expansion may also be needed to incorporate new system capabilities as they develop). A good Request for Proposal does not have to be a 100 page document, but it is important that it address key issues and be clear. Make it easy for the proposers to respond and you will get a better quality and faster response.

◼ Look For Comfortable Supplier Relationships

The company from whom you will purchase the system is likely to be the same company who will maintain the system and support its use. This is a long-term relationship. Get to know the people you'll be calling upon and find out how long each has been with the company and what his experience is. Make sure you feel comfortable with the people. If you don't, the relationship may not be a good fit. Keep looking.

Find out the average size of the supplier's customers and see how your organization fits in. If you're going to be the largest customer for a small supplier, its important to know this and set expectations for support. In general, smaller organizations tend to be happier with relationships with smaller suppliers.

Some manufacturers maintain divisions from whom you can buy their systems directly while others sell through separate suppliers only. Others combine both approaches, often in the same territory.

◼ Attend System Demos

Before arranging a demonstration of the system you will buy, talk to the supplier about how you'd like to demo set up (ideally in the precise way that you plan to use the system.) Bring your committee and take a few hours with each system under consideration. Test all system functions. Listen to the voice mail and other automated answering systems. Refer back to your list of the important system capabilities and make sure they're all demonstrated. If some of your current system's functions work well, try them out to see how the new system will compare. Don't be shy at the demo. Really kick the tires and don't let the sales person turn it into a canned presentation. While you're there ask to be shown around the suppliers' premises including the customer service department and the technician's dispatch desk. Ask how the supplier tracks repair problems and other customer requests.

Buy Locally

Always buy a system where you have options for maintenance and support nearby. The idea that a telephone system can be remotely maintained is not practical. If outside lines fail you need someone to come on site to diagnose your problem.

If the supplier is in the neighborhood, developing a good support relationship will be naturally easier for you and for them.

Understand System's Strengths And Weaknesses

Once you've addressed the system functions, evaluate the technology (*how* it makes the system functions work). Telephone system manufacturers have used a variety of approaches in putting their systems together. Some are more flexible than others. This is not a discussion to have with the salesperson so ask for a technical support person to visit and talk to your in house technical staff or consultant.

There are often tradeoffs. For example, a voice mail system that also provides fax retrieval may be appealing, but the fax capabilities may not be as flexible or complete as if you bought a separate Fax Server. But buying a separate Fax Server may require integration with the voice mail system and a relationship with a separate supplier.

Revisit the system functions that are most important to your organization to be sure you understand their operations and any limitations.

Since all systems are different, the hoped for "apples to apples" comparisons are often hard to come by, so get as close as possible but keep your energy focused on understanding the system you like best.

■ Budget For More Than You Think You'll Need

We suggest budgeting for $1,000. per user to purchase a new telephone system with voice mail. If you add other peripherals the cost may go up. If you use some of the newer breed of PBXs mentioned above, initial cost may go down. Take a close look at ongoing costs as well both for maintaining and administering the system and for keeping the system software up to date.

Another thing to be budgeted is the time it takes for this project. Make sure everyone who will be involved realizes that there will be a significant time commitment in the purchase, planning and implementation process.

We wish you success on your BIG TELECOM BUY!

Setting Up Your Telephone System to Enhance Your Business

You do want to answer those calls from people who want to give you money —
don't you?

Covering Incoming Calls With Real Live People (Part One)

"Ring, Ring...Hello?"

Since the one of main purposes of a telephone system is to enable you to answer telephone calls, it is surprising how many organizations do not think through how they are going to do this before investing in a new system. Instead, they purchase the system technology or manufacturer name, presuming that all systems are pretty much the same in terms of how they let you cover the answering of incoming telephone calls. This is not a valid presumption. Sure, every system will let you pick up the telephone and say "Hello" but Call Coverage is a lot more than that. What if you're on the telephone but waiting for that second call? What if you're away from your desk, but in the office and a customer calls to place a big order? What if five other people in your department can handle the call for you — does it make sense to send that caller to voice mail?

Call Coverage blends the functions of the telephone system with procedural considerations that may differ from one organization to another. Which calls must be answered by a live person and which can comfortably go to voice mail? How will the staff in each department back each other up for handling calls, or won't they? Are there any bosses with a secretary who screens the calls? (There are!) So by Call Coverage we mean the art and the science of getting your telephone calls answered, creating a positive impression every time someone calls any part of your company.

In this two-part tutorial we will present different call coverage scenarios and some things to consider in successfully setting them up.

If a person is sitting at his desk and is not on another telephone call coverage is simple, he just lifts the receiver and answers the call.

■ "On Hold"

During the course of the call he may need to put the call on hold. If he has other people's lines or extensions appearing on his telephone, will he be able to remember where he put the call on hold or is there some way the system can remind him (different color light, different rate of flickering on the light, etc.)?

In the same situation can someone at another telephone inadvertently cut in and speak to the person he left on hold, creating a impression of being disorganized, or is there a way to prevent this from happening?

And what does the caller hear while he is on hold? Pleasant music, information about your company and its products or just dead air — or worse yet a "hum" on the line?

What if he forgets that he left the call on hold? Does the caller just sit there forever or does the system time out and send the caller somewhere else? If so, where does the caller go? Think about the poor caller on hold who may end up in the voice mail of someone other than the person with whom he was speaking!

■ When You're At Your Desk But On Another Call

How many calls do you want to handle at the same time? In most cases, people want to see when another call is coming in for them.

Will the display on your telephone identify the incoming caller with caller ID or the caller's name when you are on another call, giving you enough time to put your first caller on hold? Do you hear any audible sound to let you know that another call is arriving and what are your options for adjusting this or turning it off? Some systems may require you to press a button to see the caller ID on the second call, which can be distracting and runs the risk that you will disconnect the first caller.

Call you tell if the second incoming call is an internal call and if so the name of the person who is calling you? Is there some way to respond back to this internal call without interrupting the call you're on — such as pressing a button to acknowledge the call?

While you are on another call, if you see a second call coming in that you have been waiting for, but cannot put the first call on hold, can you take some action to redirect that call to someone else or to voice mail?

If you see a call go into voice mail can you retrieve the caller? (It's amazing how few systems let you do this!)

Do you have others in your department who have appearances of your extension on their telephones that can answer the second call for you and keep the caller on hold until you are finished with the first call? The person can keep going in on the line to advise the caller that you will be taking their call shortly. Telephone system manufacturers have many variations on how others can cover your calls by having appearances of your extensions on their telephone. If you plan to have others cover your calls whether or not you are in the office, pay close attention to the differences in system capabilities.

When You're Away From Your Desk

What happens to callers to your telephone when you are not at your desk. Some systems make the caller wait while your telephone rings 4 times, then wait while the telephone to which you have forwarded your call rings 4 times and then wait again while the call goes to voice mail. This is not an example of good call coverage!

On the other hand, is it a good idea to have a button to press to send your callers to voice mail after your telephone rings once? Isn't it easy just to keep that button pressed even though your are at your desk, just to get some work done — as if handling the incoming telephone callers is not the most important work of the organization!

And what if others in your department can answer the call and handle the caller? Does it make sense to build up voice mail messages in your mailbox that you have to spend 3 hours the following day responding to?

The previous scenarios are the type of things to think about when planning for call coverage for a new or existing telephone system. In "Call Coverage Part Two" we'll present some options to look for that improve call coverage capabilities with a new telephone system.

Covering Incoming Calls
With Real Live People
(Part Two)

"What do you mean you weren't expecting a real live person?

In "Call Coverage Part One" we presented some things to think about in terms of how to most effectively answer your incoming telephone calls. Flexibility for setting up this call coverage is largely dependent upon the capabilities of your telephone system.

Here in Part Two we present our view of the characteristics of a telephone system that supports getting your calls answered and keeping callers happy.

These thoughts apply both to an older telephone system, to the new PC based systems and to everything in between.

Ringing (Audible Indicators)

What is more basic to the telephone than the sound of ringing? Make sure your system can adjust the ringing volume and ring differently (by the sound or the "cadence") for different types of calls. For example, you may have a customer hotline that you want to ring with a different sound, enabling those calls to be given priority. Intra-office calls can be distinguished from outside calls through ringing. You may also want a different ringing sound for each telephone in a group of people who sit together. If someone is away from his desk, he can distinguish the ringing on his telephone, increasing the likelihood that he will run back to the desk to answer it. Look for flexibility in being able to turn the ringing off and on. Some executives may want the ability to turn off the ring altogether if a secretary screens the calls. Giving everyone the ability to turn off the ring may result in a peaceful office with all callers going to voice mail or perhaps to your competitors! One last thought on ringing. When purchasing a new system, listen to your ringing options. Some systems ring with very grating sounds that you may not wish to listen to for the next 5 to 10 years.

Lights (Visual Indicators)

Some telephone system manufacturers replaced the lights with LCD type indicators a few years ago, but most of them are reverting back to lights as nothing works better to enable good call coverage. The light is typically located next to the button that you press to answer an incoming call. A new incoming call is best indicated by a light flashing on and off. When the call is answered the light becomes steady. When the call is put on hold the light flickers. There are so many people in the workforce accustomed to these conventions that work very well, that there seems little point in changing them. However, some telephone system manufacturers like to experiment with a different approach, but this usually results in confusion and a lower quality of call coverage. Even better is different color lights to indicate different call states (new call, on hold, call in progress on your telephone vs. call in progress on another telephone.) Lights are easy for most humans to see and enable them to quickly respond. Lights are also easy to see from a distance, useful since you may be near your desk, but not close enough to see an LCD display.

Hold

Since it is common for us to ask a caller to wait while we check something or look for someone, putting a caller "on hold" is an important part of call coverage and the impression we make on callers. If your telephone covers multiple outside lines or extensions, it is important that the telephone system enable you to remember on which button you left the call on hold at your telephone. This may be accomplished by a different rate of "flickering" than the rate for other calls on hold or perhaps a different colored light. A capability called "individual hold" enables you to put a call on hold at your telephone that cannot inadvertently be taken "off hold" or cut in on by someone at another telephone. Another useful capability is "hold recall" that provides a unique ringing to remind you that you've left a call on hold too long, again, particularly useful if you are answering calls on multiple outside lines or covering for other people as is typically the case in most offices or departments. Some telephones systems send a caller left on hold too long to another extension or to a switchboard operator, but this is confusing both to the caller, who doesn't know where he ended up and to the person who left the call on hold and returns to his phone to find it gone! Even worse is the system that sends callers on hold too long to a voice mail system, asking the caller to leave a message!

Buttons And Expansion Modules

A few years ago, telephone system manufacturers were paring down the number of buttons on their telephones, believing that providing information on the display of the telephone would replace the need for buttons. If you were answering a call for someone else, that person's name appeared in the display. This did not work out well and most manufacturers now realize that pressing a button with a light next to it is the fastest way to get calls answered, particularly in a busy environment where many calls may be coming in at the same time. For good call coverage, look for telephones with large well-spaced buttons and the ability to have many of them, including expansion modules that add more buttons to the telephone. Find out how many buttons are needed to cover for others in the office. Some telephone systems take up three buttons on each telephone for every other person's telephone you cover. So if an administrative person is answering for three different people, this requires using a total of 9 buttons on the secretary's telephone. Another use of buttons, in addition to covering calls for others, is to provide a visual display of who is on the telephone, sometimes called a "busy lamp field." Many managers find this to be the best way to quickly monitor activity within the office or department.

Find out if your telephone system has any limitations in terms of how many people's calls can be covered by another telephone and how many different telephones can cover for a particular extension or direct dial telephone number. If "covering" means anything but a discrete specific button on the telephone associated with each extension or direct dial number to be covered, be skeptical. If the alternative is not crystal clear to you, it may not be to others who are using the system.

Displays

As we mentioned above, although some telephone system manufacturers tout the display on the telephone as the solution for good call coverage, the reality is that most people either cannot see the display or ignore it, so this has not really worked. Most displays are indeed hard to see. Lucent Technologies (now renamed Avaya) once had a beautiful display with characters in a readable bright blue, but they discontinued it. Many display messages are written in cryptic telephonese such as TK 001 or XFER VM that means little to the user. Spend some time setting up the messages that do appear on your displays, making them meaningful and easy to understand and you will automatically improve call coverage. A good application for displays is Caller ID, enabling you to see who is calling from the outside. Since telephones do now come with displays, look for a large one that is adjustable, can tilt and that is easy to read in any light, including sunlight coming though the window, typical in many offices.

Call Forwarding

Ideally, incoming telephone calls are answered by someone who can help the caller. If this is not possible, many telephone systems have capabilities to "forward" calls somewhere else. The "somewhere else" may vary, depending upon the circumstances. The called person may be unable to answer because he is busy on another call. In this case he may want a new incoming call to be forwarded to another button on his telephone so that he can put the first call on hold and answer the second call. (This is sometimes called "hunting" or "rolling over" rather than forwarding.) Or he may not wish to be interrupted and may want new calls to go to his voice mail when he is on another call. A third option may be for calls to be forwarded to another telephone where a back up person can take the call for him. A fourth option would be for the caller to hear a busy signal. Some organizations prefer this, particularly for intra-office calls. The busy signal quickly tells the caller that the person she is calling is there, but on the phone. If this call is sent to voice mail the caller has less information in terms of whether the person she is trying to reach is actually there . If the called person is out of the office, the forwarding scenario may be different. He may want the calls to go directly to voice mail without making the caller wait while his telephone rings. Or he may want the calls to forward to his cell phone so he can cover calls on the road. The best telephone systems provide flexibility in call forwarding. Look for systems that provide at least four different destinations for forwarding calls depending upon the circumstances. For example: (1) intra-office call if person is there but on another call — caller hears a busy signal and knows to call back (2) intra-office call if person is out of the office forwards to the cell phone of the called person (3) outside call if person is there but on another call forwards to the another button on the telephone so it can be answered. (4) outside call if the person called is out of the office forwards to voice mail for the caller to leave a message.

While no telephone system has it all, some are better than others in providing the type of call coverage that we've seen work best.

You will notice that our Call Coverage tutorials did not focus on automated systems such as Voice Mail. We observe that live answering does the most for providing a good impression to callers. In "Call Coverage with Voice Mail" we'll demonstrate how the automated systems can best complement live answering and fit into your strategy for good call coverage.

Using Voice Mail and Automated Answering to Support Your Call Coverage Plan

Let's get it right with automation!

"Call Coverage Parts One and Two" focused on live answering by a real person! Practically speaking, organizations rely on automated systems to do some of the Call Coverage when the office is closed or when no person is available to answer. Below are some suggestions for keeping it friendly and easy to use, minimizing the caller frustration typically associated with automating the answering of telephone calls.

▨ Automated Attendant

Let's start out with the Automated Attendant. When you call into an organization and hear "Thank you for calling. If you know the extension of the person you want to reach, you may dial it now or for a directory of names press 4" you are hearing an Automated Attendant. Some Automated Attendants provide options such as "for Sales press 1, for Customer Service press 2, etc." While an Automated Attendant can perform the same function as a human in answering calls and directing them to an extension, it will never replace the warmth of a pleasant, knowledgeable person in terms of the impression made on callers. If you do decide to use an Automated Attendant as the first thing your callers hear, here are some suggestions to improve the experience for them.

1. If you have a small office, don't make the caller go through finding a person's extension in the automated directory. Instead, put the people's names right up front as selection options: "For Jane Laino press 1, for Diane Ventimiglia press 2, for Jean Fitzpatrick press 3, for Laura Taylor press 4." In a slightly larger

organization, you can put the most frequently called people up front and then say, "For a directory of other names press 8."

2. If you must send a caller who does not know the extension he is calling to an automated directory, make sure the directory is accurate and is kept current. Nothing is more frustrating than trying to reach someone you know is there, but not finding the name in the directory. Try to set up your directory so that callers who misspell names or may only know the person's first or last name can still get the extension.

3. Investigate the Automated Attendants that enable the caller to speak the name of the person he is calling. This uses what is called "Speech Recognition" technology. This is the closest thing to live answering, as the Automated System can be spoken to by the caller. The system then directs the caller to the extension of the person he requested.

4. Avoid multiple levels of selection options on an Automated Attendant if possible. For example, once the caller has "pressed 1 for sales" he is then presented with other options such as "press 1 for tire sales, 2 for auto parts sales, etc. " If you must do this, keep it to two levels. Three is just asking too much of the caller's patience. Carefully think through the options being presented, to be sure they coincide with caller needs. Get feedback from callers in terms of how easy the system is to use and monitor actual use through the reporting capabilities of the system.

5. Always enable callers who just "hold on" or press "O" to reach a live person who will know that the caller has heard the Automated Attendant and has encountered difficulty or elected not to use it.

Voice Mail

Whether you have successfully navigated through the Automated Attendant, gone through a live switchboard operator or dialed directly to a person's desk, if that person cannot answer you may hear an recording of that person's voice saying, "This is Christine Kern at CMP Books, please leave a message and I will call you back." You may then leave a message that is recorded by the system and can be retrieved by the person you called. That's Voice Mail.

Some suggestions for successfully using Voice Mail to support Call Coverage include:

1. Establish a company or departmental policy and procedure for using voice mail and stick to it. Some organizations decide to answer all calls live, then offer to transfer the caller to voice mail if the called person cannot be reached and the caller wishes to leave a message. Some may also offer the caller the option leaving a message with the person who answered, which is always good form if time permits. While this does take more manpower, it is considered by many to be a kinder approach than just dumping the caller right into the voice mail announcement.

2. If someone calls into an extension and goes directly to a voice mail announcement, give him an option of pressing "O" to escape to a live person. Since it has become something of a convention for people to dial "O" (for Operator) to reach a live person, it does not make sense to have a voice mail system that uses any other digit or series of a digits and letters along with the * or # sign (the worst!).

3. Some voice mail announcements instruct the caller to press another extension number to "reach my assistant." When this happens, it is typical to get the voice mail of the assistant! Now the poor caller cannot even leave a message for the person he was calling since there are never instructions for how to go back to the called person's voice mail. Having a caller escape from one person's voice mail and end up in another's is yet another form of caller abuse.

4. In terms of "escaping" from voice mail. We do recommend that callers pressing "O" not be sent to a switchboard attendant who has no idea of what is going on in the department, but rather to a departmental extension that is always answered. To avoid relying on just one person for this back up, designate a separate button on everyone's telephone that, when it rings, they will know that someone has exited voice mail and wishes to speak to a live person. The display on the telephone will indicate whose voice mail the caller encountered, so that the call can be answered appropriately. In the case where you must send voice mail escapees back to the switchboard, be sure that the switchboard attendant can identify that this caller has been in voice mail and can give the call priority.

5. Develop a standard for the scripting of individual voice mail greetings within your organization. Don't leave this to chance or callers will get an impression of inconsistency, depending upon who is called. We recommend that the voice mail greeting include the name of the company. "This is John Prescott at Comware Systems, please leave a message and I will call you back." Remember, a caller may not personally know the individual he is calling. Suppose someone calls a direct dial number and an unfamiliar name answers (this often happens when one person forwards his calls to another person for answering). If the person's greeting includes the company name, the caller will at least know he reached the right organization.

6. Avoid overused voice mail greeting phrases such as "I'm either on the phone or away from my desk." Some voice mail systems can determine whether you are on the telephone and, if so, will give the caller the option of pressing 1 to wait for you or pressing 2 to leave a message.

7. Decide on a company policy in terms of how long it takes to return voice mail messages. Avoid stating it in the voice mail greeting however, as it may create an expectation you cannot

always live up to such as, "I will return your call within 4 business hours.

8. In general, we suggest that you not say too much in a voice mail greeting (taking up the caller's time) or change the greeting very often. Invariably someone will be calling you on Wednesday and your greeting will say it is Tuesday, making you and your company appear disorganized.

9. Train your staff on how the telephone system is set up to work with the voice mail system. Most people do not know the different circumstances under which callers are sent to voice mail. Also train staff on how to transfer callers into the voice mailboxes of others. Very few people know how to do this properly, resulting in callers having to wait too long or hear an inappropriate greeting such as "Please enter your password."

10. Don't program your telephone system so that callers are only sent to voice mail when you press a button on your telephone. If someone forgets to press the button, calls may never be answered. The button can be used to send callers to voice mail without their having to wait for your telephone to ring, if you are truly not at your desk. If you forget to press the button, callers will wait longer, but will still reach voice mail if you set it up properly.

Voice Mail and Automated Attendant are typically capabilities of the same system. Recognize that in most cases this is a separate system from the telephone system (PBX) with separate care and feeding requirements. Arrange for regular maintenance and review reports on system operation to ensure that it continues to provide good call coverage support that you have put so much time and energy into planning.

Ten Simple Tests for Your Automated Attendant

*"For instant gratification,
if you want to press one — press one!"*

"Thank you for calling The Very-Busy Company. If you know the extension of the person you are calling, you may dial it now. For Sales press 1, for Customer Service press 2 or press 4 for the company directory."

Does your organization use an Automated Attendant that sounds like this to answer your main telephone number? If so, why not try these simple tests below to see if you're making it easy for your callers.

Remember there are three categories of callers (1) people who are not accustomed to using Automated Attendants at all (2) people who are accustomed to using other Automated Attendants who have some preconceived impressions of how to navigate through them and (3) people who call you regularly, know how your system works and want to move through it quickly. A well-designed system must accommodate all three types of callers.

The objective of the Automated Attendant is to get callers to the intended person or department quickly and to leave the caller with a favorable impression of your company. If your system is hard to use you are sending the message that doing business with your company is hard, too!

While the Automated Attendant is often a capability of the system that also controls Voice Mail, it is a separate function. The Automated Attendant directs a caller to his destination, while Voice Mail takes messages.

So sit down, grab a pad and pencil to take notes on what you find and start dialing…

■ TEST #1 — No waiting for a caller who knows the extension of the person he is calling

If a caller knows how to use the system, he does not want to wait to listen to an announcement. Call into your main telephone number. When you are answered by the Automated Attendant, using the touchtone buttons on your telephone and without waiting, dial the extension number of a person in your company. In the best systems you immediately hear ringing, indicating that the telephone extension you're trying to reach is ringing at the person's desk.

■ TEST # 2 — No waiting for a caller who knows the selection number he wants from your automated attendant announcement

If your Automated Attendant offers options such as: "for Sales press 1", callers who have used the system before will not want to wait to listen to an announcement. Call into your system to see if you can press 1 as soon as the call is answered, without waiting to listen to announcement.

■ TEST #3 — Immediate escape capability by dialing "O"

Like it or not, many people simply do not wish to use your Automated Attendant. For the best service, callers who dial "0" (zero) are immediately sent to a live attendant.

■ TEST #4 — No critical instructions for using your system are left until the end of an announcement. The caller may not know to wait for them.

Many Automated Attendant systems require that you "press the pound sign (#)" after dialing an extension or entering the first three letters of the person's last name to use the company directory. If a caller who uses other Automated Attendants encounters yours, he may start pressing touchtone buttons and not wait to listen to the end of the announcement. In this case the caller's request will not be processed. If you do discover that this occurs in your system, find out what happens to the caller when he fails to enter the "#." Try to get the message to provide this information right up front such as, "Press # after entering the extension of the person you are calling."

TEST #5— Flexible "spell-by-name" function of automated attendant directory. Not limiting to the caller to three letters

If your Automated Attendant announcement offers the option of a company directory (and most systems do), call into your system and select the directory option. A typical system may direct you to "enter the first three letters of the last name of the person you are calling." The best systems do not limit you to three letters, but let you keep entering letters until the name is recognized. If your system responds to only three letters, what happens to callers who enter 4 or more, perhaps not waiting to hear the part of the announcement that asks for only three?

TEST #6 — No first names in the "spell-by-name" directory

Setting your system up so that it conforms with how the majority of other systems are set up is a good idea. Companies who decide to use "first names" in the company directory with the idea that it sounds more friendly often make things difficult for callers. A caller may just wait to hear "Enter the first three letters of ..." and then enter the person's last name which is how most systems work. In addition, it is more likely that people have the same first name that the same last name, so using first name may decrease the likelihood that the caller will get to the right person.

TEST #7 – Put the names of the most frequently called people at the front of the list offered by the directory

Some Automated Attendant directories, upon recognizing a name you have entered, offer a list of people whose names match. For example, "Brenda Jones. If this is correct press #, if it is not correct press 1." The caller presses "1" and then hears "Michael Jones, if this is correct press # etc." If Michael Jones gets many more calls than Brenda, why make more callers wait to get to his name? Put the most frequently called people at the front of the list offered by the directory.

TEST #8 — Give the caller the person's extension number for future use

A caller using your Automated Attendant directory may want to jot down the person's extension. Make sure your directory does give the caller the extension number. "Your call is being transferred to Michael Berezein, extension 103." Some systems do not.

TEST #9 — Make it clear what the callers is to do, once he gets the extension number from the automated directory

This is one area where different manufacturer's Automated Attendants are not consistent in the way they work. For example, after the caller has successfully spelled the person's last name (or part of it) using the touchtone buttons on his telephone, what is he supposed to do next? Some systems will say "Megan Coe, extension 202" and expect the caller to know that he now has to press "202" while other systems will say "I will dial that extension for you now" requiring the caller to do nothing. Still other systems will provide the instruction to press "#" or press "1" to direct your call to that person once you have heard the name.

■ TEST #10 — Pretend you are a "neophyte" caller

While many of us have become somewhat savvy in figuring out how to use Automated Attendants, many others have not. Try to pretend that you have never encountered such a system before and carefully listen to the instructions given at each juncture to see if you can follow them. Even better, get someone who truly is a neophyte to try it. Try dialing a number that is not offered as a selection. Try dialing 5 digits when your system answers even though your company extensions use 4 digits. Try doing nothing when answered and see how long it takes for you to be answered by a live person. Try entering two letters of the person's name in the directory when you are instructed to enter three letters. Do just the opposite of what the instructions tell you to do and see what happens. The best systems will send such a caller to a live person for help. The worst systems will announce "Thank You for Calling. Good-bye!!" This is exactly what many systems do. Be sure yours isn't one of them!

We suggest conducting this test both during the day and when your office is closed. When the office is closed and there may not be a live person available, the announcements heard need to reflect this. If a caller presses "0" and hears that his call is being transferred to an attendant, it is important that one be there. (Test #11!) At night this announcement should be changed to indicate that the office is closed.

You will undoubtedly discover other things about your Automated Attendant while conducting this test. Just jot them down, type up your list and review it with your system administrator or the company who provides service for your system.

Ten Simple Tests for Your Voice Mail System

"I'm either on the phone or away from my desk — but it's up to you to figure it out!"

Here are ten simple tests to see how well your Voice Mail system reflects on your organization. Voice Mail is the system that enables a caller to leave a message for a specific person in that person's "Voice Mailbox." Try these tests to find out what kind of impression you're making.

TEST #1 — What do callers hear while waiting for Voice Mail?

Dial your organization's main telephone number and wait for the Automated Attendant to answer. ("If you know the extension of the person you are calling you may dial it now, for sales press 1, etc.") If you have a live switchboard attendant during the day, conduct this test after hours. Upon reaching the Automated Attendant, dial the individual's extension number and make note of what you hear next. In the best systems, the caller hears ringing (this gives him the most appropriate reinforcement that his call is being sent to the extension he requested). If no ringing is heard, at least pleasant music or an announcement may be played. The worst scenario is that the caller thinks he has been disconnected when he encounters "dead air," hearing nothing.

TEST #2 — How long do callers wait to reach Voice Mail?

Dial into your Automated Attendant again, then dial the individual's extension number. Note how long it takes for the voice mail greeting to answer. The shorter the time, the better. In the best systems, the caller will not have a perception that he has waited at all (5 seconds or less). In the worst systems the caller may wait 15 to 30 seconds or more. The length of this wait may be a variable in your telephone system that can be adjusted. With many telephone systems, when a person leaves his desk during the day or for the evening, if he presses a button on his telephone that "forwards calls to voice mail" the caller will not have to wait as long to hear his voice mail greeting. People often forget to press this button. Find out how your system works and make sure that the people in your organization understand it.

■ TEST #3 — Do individual voice mail greetings that can be reached by dialing a unique telephone number identify your organization's name?

Listen to an individual's voice mail greeting. If you can reach a person's voice mailbox by dialing his own unique telephone number (called a DID or direct inward dial number), it is important that his greeting include the name of the organization. "This is Aldo Forlini with Focal Communications, please leave a message and I will call you back." Otherwise, if the caller is responding to a message from an individual he does not know (a salesperson, for example), he may be confused as to what organization he has called.

■ TEST #4 — Do individual Voice Mail greetings in your organization use tired, overworked phrases?

Listen to the individual greeting again. Better yet, listen to five or ten, to get a reasonable sampling. Be sure that your organization does not use tired phrases such as "I'm either on the phone or away from my desk" or even worse, "Please stop and listen to this message" as if the caller would be doing anything else at that moment. It is important that your organization set a policy in terms of how individual voice mail greetings are recorded so that they will mirror the manner in which you wish to be perceived. "This is

Stacey Maites at Avaya. You have reached my multimedia mailbox" appropriately reflects that a technology company is making use of the technology it is selling. Or "This is Laura Taylor at DIgby 4 Group, please leave a message and I will call you back." Simple and to the point. Don't make a busy caller listen to a long-winded voice mail greeting. Listen to your own voice mail message along with others and see how you rate them.

TEST #5 — Does the unsuspecting caller also reach the Voice Mail of the person who is covering for you?

Does your voice mail greeting give the caller the option of deciding not to leave you a message, but instead be sent to a "covering extension" (someone who is taking calls for you). If so, it is critical that a real person be there at the other extension at all times to answer the telephone. Once the caller has encountered the second person's Voice Mail greeting, there is never an easy way for him to get back to your voice mail to leave a message. This is highly frustrating. Call into your system and opt for the covering extension. If you reach the Voice Mail of the person who is supposed to be backing you up, your organization has failed this test!

TEST #6 — Can a caller easily exit from your Voice Mailbox and dial someone else within your company?

Can a caller who knows the extension numbers of others in your company dial in, reach your Voice Mail greeting and then decide to try another extension rather than leaving you a message? How is this done? How are the instructions given to a caller who may know another extension, but does not know how to get to it while listening to your Voice Mail greeting? Dial into your Voice Mailbox and then try to get to another extension number. A good system will allow you to do this and make it clear how to do it. Dial in again and pretend that you do not know another extension to call, but have the name of another person and need to get to a company directory. Can you do it?

■ TEST #7 — Does your Voice Mail system require callers to dial unfamiliar strings of symbols, letters or numbers to reach the switchboard attendant?

Call into your own Voice Mail greeting and listen to the instructions given to callers to reach "an operator". In a well-designed system, if the caller presses "O" (for operator) at any point, he will be sent to a real person (either the switchboard attendant or a back up person). Since most people familiar with Voice Mail systems expect to be able to dial "O," systems that require you to dial something else such as "*T222" or "#O" to reach a real person can be confusing to callers.

■ TEST #8 — After you retrieve your messages from your own Voice Mailbox, can you easily leave a message for someone else in your organization?

Dial in to retrieve your own voice mail messages. When you have finished, can you easily get to the Voice Mailbox of another person within your organization to leave a message? If you don't know how to do this, does your system offer any helpful messages on how to accomplish it? In the worst systems, you'll have to hang up and call in again to leave a message for someone else.

■ TEST #9 — Don't use your Voice Mailbox as a storage space for messages and let callers encounter the announcement "This Mailbox is full!"

Call into a number of different Voice Mailboxes within your organization to see if you reach any "full mailbox" announcements.

This is very frustrating to callers. A full mailbox may be the result of someone storing and saving many messages or it may be that your particular system does not have sufficient disk space. Most voice mail systems delete old messages after a certain period of time, which is adjustable. Most Voice Mail systems also have a variable parameter in terms of how many messages can be stored before the "full mailbox" announcement is heard. Make sure the people using your system know what this limit is.

■ TEST #10 — Does your system detect when the caller is finished leaving a message and then provide him with options?

The best voice mail systems detect when the caller has stopped speaking and assumes he is finished leaving his message. Then the system presents options such as listening to his message, appending to it, deleting and re-recording it, marking it urgent, etc. Some systems tell the caller to press "#" or some other button on the telephone after leaving his message, but if this direction is given before the caller leaves the message, it is not likely that he will remember the instruction after he has finished speaking. Many systems leave the caller waiting indefinitely after he leaves the message, waiting to hear what his options will be.

With all Voice Mail capabilities ask yourself three questions about your system:

1. Can the system do this?
2. Is it easy to do?
3. How does the system instruct callers how to do it?

We hope that your Voice Mail system has proven to be friendly to callers. If not, meet with the company who maintains the system for you to see what you can do about improving it.

Using Your "On Hold" Capability Effectively

"I'm so sorry, I forgot I left you on hold while I went to lunch."

The act of placing a caller "on hold" has become second nature to each of us working in the office environment. With the exception of Incoming Call Centers where customer service is a focus, it is a rare organization that thinks about its procedures for putting callers on hold. Yet this is a critical element in providing good service and creating a positive impression of your company.

Placing a caller on hold enables you to temporarily leave the caller without disconnecting the call. The caller remains at your telephone, but cannot hear you after you have pressed the hold button. Pressing the button where you left the caller enables you to retrieve the call.

To think about your procedures, a good starting place is to understand how the capabilities of your office telephone system work as they relate to using this "hold" function. Keep in mind that every telephone system works differently.

Can You Find the Hold Button?

Placement

Where is the Hold button on your telephone? It is important that it be easy to reach, easy to find and not right next to other buttons that look exactly the same. On most telephones pressing the wrong button when trying to put someone on hold results in the caller being disconnected! Many telephones are designed with a button called "Release" that hangs up on the caller and is regularly mistaken for the Hold button.

● **Size**

How large is the hold button? Ideally it is larger than the other buttons to make it easier to locate and to reduce the likelihood that you will press a nearby button in error.

● **Color**

Make sure your telephones have a hold button that is red or orange, making it easier to locate. Trying to read the small print that telephone manufacturers use to label the buttons can be difficult, especially when you're in a busy office with many things vying for your attention! If your telephone has only clear plastic buttons, make your own colored label to distinguish the hold button from the others.

■ Can You Find the Caller You Left On Hold?

If your telephone has appearances of multiple outside lines or extensions, often shared with others in the office, how will you remember where you left a caller on hold?

● **Color of the light**

On most telephones there is a light associated with each button on which a call may be in progress. On many systems, new incoming calls have a green light. When the call is answered, the light turns to red and when the call is put on hold, the light remains red.

● **Rate of "flickering"**

On most telephones, when a call is ringing, the light next to the call has a pattern of "on and off, on and off, etc. " until the call is answered. Once the call is answered, the light remains lit. When that call is placed on hold, the light "flickers" (going on and off in rapid succession). On the better telephone systems, the rate of flickering for a call you put on hold at your telephone is different from that of the calls put on hold by other telephones (even thought those outside lines or extensions may appear on your telephone.) Getting to know the appearance of the lights for different conditions on your particular telephone system is important. (The behavior of the light is also know as "flashing" and its speed known as the "rate of flash." These terms come in handy when trying to communicate with your telephone installation or repair person.)

Individual Hold

Some telephones enable you to place a caller on hold in a manner that ensures no one else in the office who picks up on that same outside line or extension will be able to take the caller "off of hold." This is called Individual Hold (also sometimes called I-hold or Exclusive Hold.) On some systems this is accomplished by pressing the hold button twice, but find out how your particular system operates.

What Happens when You Leave a Call On-Hold Too Long?

Most telephone systems have a function known as "Hold Recall." This means that when you've left a caller on hold too long at your telephone, the telephone will start ringing (with a different ringing than a new incoming call) and the light where you left the caller on hold will flash. This is a reminder. You can set up your telephone system to determine what length of time will pass before Hold Recall starts. We'd suggest setting this up for no longer than two minutes.

Some telephone systems also have the capability to send a caller who is left on hold too long back to the main switchboard attendant. We do not recommend this, as it is confusing to both the caller ("Where did I end up?") and the switchboard attendant who typically has no way of knowing that this was a caller left on hold too long. Make sure that your telephone system does not have this function enabled.

Faster Ways to Put Callers On Hold in High Call Volume Environments

Telephone system manufacturers have come up with several alternatives to having to find and press the Hold Button to put a caller on hold. These are sometimes implemented in departments where there is a high volume of calls.

▶ "Pressing the Button Where the New Incoming Call is Ringing" *twice* can put the caller on hold in many systems (be careful, since in other systems, doing this will disconnect the caller!)

▶ "Automatic Hold" means that when you press a button to answer a second incoming call, the first call will automatically be put on hold.

These and other changes to the more conventional pressing of the hold button can indeed save time, but must only be used by people who know how they operate.

Using the "Call Park" Function in Place of "Hold"

Suppose you wish to place a caller "on hold" and go to another telephone in the office to retrieve the caller. (maybe you are in an open area and wish to take the call in a private conference room). Unless that other telephone happens to have an appearance of the same outside line or extension that the call is on, you may use the "Parking" function of your telephone system. Again, systems operate differently, but (for example) you may press a button on your telephone that is labeled "park" then dial in a two digit "parking code" and hang up. The call will disappear from your telephone. You now have a pre-determined amount of time to go to the other telephone, lift the receiver, dial the two digit code and retrieve the caller you "parked." If you do not retrieve a parked call, the caller may be automatically redirected to the switchboard attendant.

In some offices the switchboard operator parks calls and uses the overhead paging system to announce, "William Cockrell, you have a call on 24." Then Will can go to any telephone in the office to retrieve the call.

In some systems, callers who are parked may not hear the "on hold" message or music while waiting. Find out how your system works.

What Does Your Caller Hear While On Hold?

Most telephone systems have the capability for "Music On Hold" which can actually be music or some other type of announcement. If you decide to use music, make sure that it is something that your callers will enjoy. We suggest tapes or CDs rather than having an FM radio tuner as some organizations do. You run the risk of having callers listening to advertisements from your competitors or listening to disconcerting news reports. Pleasant music is much better and keeps the caller in a more relaxed frame of mind while he waits.

There are also services that develop customized on hold messages combining music and professional announcers talking about the services of your company. Many of these "Message On Hold" companies provide regular updates so that you can keep your message current in terms of new products and promotions.

Since what callers hear while on hold is a reflection of your business, give some thought to the impression you want to convey. If your company is a white shoe professional services firm it may be jarring for callers to hear the Grateful Dead when put on hold. But if you're a "downtown" movie production studio this might be just the right thing!

In any case, no matter how entertaining the on hold message, your callers have better things to do, so keep this waiting time to a minimum.

Protocol for Putting Callers On Hold

Think about the impression you are giving to your callers. If you answer "Phil's Hot Dog Stand, please hold" and then press the hold button without giving the caller a chance to speak, this creates a very poor impression.

While it may take a few seconds longer, answering "Phil's Hot Dog Stand, will you wait just a minute please?" and giving the caller time to say, "Yes" is much better.

The term "on hold" in itself is not pleasant to callers so we'd suggest replacing "hold" with "wait", always asking the caller for his agreement to keep him waiting.

It is also good form to offer the caller the option of waiting or being called back if he is going to be waiting for any length of time.

You may also to set up a time limit beyond which your organization does not wish to keep callers on hold and make sure that your staff knows about it. Another related consideration is that if your callers are reaching you on a toll free number (like an 800 number) you are paying for each minute that caller waits!

Today's average business telephone system provides no reports on how long each caller was left on hold or how many times the same caller was placed on hold in the course of the conversation. This type of information would be of interest to those managing the organization, to provide insight into the telephone activity in each department or for each individual.

Hopefully when the next generation of telephone systems under development reaches maturity, this and similar types of useful information will be readily available.

Setting Up A Small Incoming Call Center

Find Out How Long Your Callers Are Willing To Wait!

This tutorial will provide you with guidance in setting up a small Incoming Call Center. Any company or department within an organization handling a steady stream of telephone calls may be considered a Call Center. The first step may be to define it as such. You may have been referring to it as your Customer Service Department or Help Desk or Dispatcher, but looking at it as a Call Center is the first step in expanding your thinking in terms of putting a better call handling/transaction processing system in place.

Identify Processes And Transactions First

Before you begin to think about the telephone system, take the time to step through each of the processes and transactions that will take place in your Call Center. The activities that accompany telephone calls become such second nature that, unless they are intentionally observed and documented, needed capabilities and opportunities for improvement and automation may be overlooked. This applies whether you are starting the Call Center from scratch or revamping an existing department. Write this all down, make flow charts, get a consensus from management and anyone who will be supervising the Call Center.

Put A Dollar Value On Each Call

Try to assign a dollar value to the average call. This will ultimately help you to justify the expense of setting up the Call Center equipment and services. It will also force you to think about the types of calls that come in and their significance to your organization. Calls that result in a sale may be easiest to value. What about calls to cancel a sale or change an order? The cost of this needs to be considered in terms of how many calls are actually handled to complete a sale. If calls are for customer service, the value of the call may be in the retention of the customer. Whatever method you decide upon to come up with the number, get concurrence of Upper Management early on. That way, if the new Call Center enables you to handle more calls, your cost justification will be easier.

◼ Identify Your Callers And How You'd Like Them To Be Handled

Who are the callers? Do you want to handle each of them in exactly the same way? Do you want calls from existing customers to have priority over potential new business calls? Do you want new business to have priority over customers calling with questions about existing orders? Do you represent different product lines and want callers to hear different messages while they are waiting for a customer service representative? Must all callers be handled by a person, or can automated systems take some of the more routine calls. Telephone systems for Call Centers are becoming increasingly adept at helping you to handle different types of calls in different ways. There are several different methods for finding out who the callers is or what he wants as the call arrives.

1. Caller ID or Automatic Number Identification identifies the caller by enabling your system to capture the telephone number from which the person is calling. This must be cross-referenced with a database you have already established, linking that caller's information to a particular calling number. This tends to work best if your callers always call from home or from the exact same telephone number.

2. DNIS (Dialed Number Identification Service) enables you to have many different 800 numbers to give out for different types

of calls. You may order these groups of 800 numbers from your long distance carrier. This can make a small Call Center seem a lot bigger! Your telephone system recognizes the 800 number dialed and routes the call accordingly. You can even have different types of calls going to the same person, but display on the telephone or whisper the type of call into the representative's ear so they can answer appropriately. If you do not use 800 numbers, you can get the same effect by ordering Direct Inward Dial (DID) telephone numbers from your local telephone company.

3. A more foolproof way of identifying existing customers as the call arrives is to assign each a customer I.D. number which they will be prompted to enter using touch-tone signals. Voice recognition systems are also improving dramatically, which will enable the caller to state their name and your system will recognize it and route the call accordingly.

4. Menu options presented to your callers further refine the purpose of the call and can give your callers a choice of how they want to be handled. For example, "For new orders press 1 or to check on an existing order press 2." Or "There are two callers ahead of you, if you want to continue to wait press 1, to leave a voice mail message press 2 or to use our automated ordering system press 3."

What Information Will Help You To Manage The Call Center

The Call Center telephone system can provide you with a wealth of information to assist you with managing the systems, the representatives and the level of service experienced by the callers.

▷ Real-time Information for Customer Service Representatives: How many callers are waiting to be answered? How many voice mail messages have been left? How long have I been on this call? What type of call is coming in right now?

▶ Real-time Information for Call Center Supervisors: Which representatives are waiting to receive calls? Which reps are closed and should be available? Which reps are not adhering to their schedule of breaks?

▶ Real-time Information for Managers: How many calls have come into the center today so far? How much money has been generated by those calls? How long did the average caller wait to be answered? How many callers chose to use the automated system rather than wait for a representative?

▶ Historical reports look at the same type of information as real time, providing the perspective of the way things were handled over time. They can also evaluate each representative's performance.

When you are shopping for a Call Center telephone system, the more detail it can provide about what callers are encountering, the better. Some systems will let you track caller experiences on a call by call basis. For example, if you get a complaint that a caller could not get through to a representative at 4:50 P.M. you can look at calls that came in around that time to see what may have happened.

Selecting A Telephone System

Now that you've thought through how you want to operate, the fun begins. You will now try to match up the Call Center telephone system with your expectations. The first thing to remember is that this will never be a perfect match. Expect a lot of small surprises and a few large ones. But forge ahead - the end result will be worth it!

Call Center Telephone Systems are most typically what is called an Automatic Call Distribution system also known as an ACD. This type of system was pioneered by Rockwell back in the '60s to work with airline reservation centers handling a large volume of calls. The average office telephone system has approximately 1 outside telephone line for every 4 people. The ACD may have

4 outside telephone lines for every person, the theory being that callers will hold on for a certain period of time, waiting for a representative. To be sure that the callers wait the shortest time possible, the ACD equitably distributes the next caller in line to the next available representative. It stacks up (also called queuing) the waiting callers, playing them delay announcements, perhaps giving them options for leaving messages or using automated systems and collecting statistics on all of this activity. This enables Call Center managers to appropriately staff the Call Center to provide a high level of service, whatever that is determined to be. People waiting for a help line for software may wait for an hour. Callers wanting to place an order with options to shop elsewhere may wait no more than 2 minutes.

As the Call Center concept has trickled down to smaller and smaller groups of people, manufacturers have developed systems to accommodate this environment. If your entire operation is a Call Center, you may want to look at the "Standalone ACD." If you already have a telephone system and want to set up a Call Center for one department, you may want to investigate whether your office PBX can be set up with an ACD as a subsystem (you'll need to purchase more hardware and software for this.) Some smaller systems may be set up as Call Centers using supplemental software from 3rd parties. The advantage of the standalone ACD suppliers is that they are clearly in the Call Center business and understand it the best, but the systems are not for everyone. One downside of ACDs is that they are primarily designed for representatives to handle one caller at a time. If your environment requires the juggling of multiple simultaneous calls by a single person, this may not be for you.

Put together a Statement of Requirements which describes, in English, exactly how you want the calls to be handled. Draw flow charts to illustrate your expectations. Order the exact number and type of telephones and capability to handle outside lines. Find out the cost for expansion and the cost for maintenance. Get proposals in writing and ask about anything that is not clear. Check references and vendor experiences with environments such as yours.

The addition of computer intelligence to telephone calls is enhancing the capability of the Call Center. For example, when a call comes in from an existing customer, a screen of information about that caller (called a "Screen Pop") can arrive at the representative's desk along with the call. Many systems are just now beginning to be deployed using a computer server as the control for the telephone system.

What Will Callers Hear When They're Not Talking To A Representative?

Focusing on the announcement capabilities of the telephone system is critical. What will callers hear when they are waiting? Don't let them hear the same grating recording over and over again if they are waiting 20 minutes. Think about the impression you want to convey with these announcements as much as you think about how the reps will handle the callers. Play music, give the callers options to listen to informational recordings while they are waiting and entertain them. If you're going to tell them how long they'll be waiting, make sure this information is accurate. It may be more accurate in the small Call Center to tell the caller how many other calls are in front of him.

Will voice mail be incorporated into your Call Center design? How will this interact with the ACD? Be careful that callers going into voice mail, then coming back into the telephone system are not counted as two callers (this happens all the time!). Think about using automated systems such as Interactive Voice Response (IVR) enabling callers to check (for example) order status or obtain account information without waiting for a representative. Don't waste valuable representative time giving out routine information such as your hours of operation or directions to your office. All of this can be prerecorded and provided as an option to callers. A Fax Server can also enable callers to receive faxed information on products and services without the help of a representative.

Some systems enable representatives to prerecord their voice so that they do not have to say over and over "Good Morning, this is Esmerelda, how may I help you?" This is called "Agent Announcement" capability, available with some systems. And keep the greetings short. It is really not good service to say, "Good Morning, ABC Company, this is John speaking, how may I be of service?" Whew!

Ordering Outside Lines

As with any telephone system, your Call Center will need outside lines to enable callers to reach you. You may order 800 numbers from your long distance service provider either over a high capacity circuit known as a T1 (each T1 handles 24 calls at a time and can handle both incoming and outgoing calls) or regular telephone lines (one call handled per line). You will also need some lines from your local telephone company for receiving and placing local calls. The important thing to remember here is that you can only have as many callers as you have outside lines to handle them. Start out by having more lines that you think you need. Don't sign any long-term contracts for a few months until you find out what you need. If you do, get a clause that will let you out if it turns out you've over ordered.

"Right-Sizing" Your Telephone System

Not too big, not to small

Did you ever...

▷ Buy a telephone system with more capacity than you will ever need?

▷ Buy a telephone system that is too small, so you had to replace it?

▷ Buy too many connections from your telephone system to the outside world, paying monthly charges for lines you never use?

▷ Buy an insufficient number of connections from your telephone system to the outside world, so that your customers reach a busy signal when they call you?

Preventing these is what this tutorial is all about.

Like many aspects of telecommunications system management, the concepts of scaling and trunking affect organizations in three areas:

▷ Maintaining Service Levels for Callers and Staff

▷ Controlling Expenses

▷ Remaining Flexible to Accommodate Changing Business Requirements

Definitions

The term *Scaling* as we are using it here to relate to the PBX and Voice Mail means making sure that the system hardware and software components are the right size and that you are buying the correct number of each of the system components. It also means that you are buying enough room for growth. The objective is to ensure that the system will work in the way you expect it to both when you buy it and for the life of the system. A further objective is that you not spend more than necessary.

Trunking is a term relating to the outside lines that connect your telephone system to the outside world. Sometimes these outside lines are called trunks. The concept of scaling also applies to these trunks, making sure that you have enough, but not too many and room for growth.

An Overview

Our approach to this topic will be to identify each of the items that need to be addressed in the scaling or sizing of a PBX and of Voice Mail.

Next, we will talk about trunking and how to determine how many trunks you need, starting with the dartboard method up through using statistical tables and computer modeling software.

Scaling The PBX

In order to think about the sizing of a PBX, it helps to visualize how one is typically put together. It may be a big refrigerator sized cabinet or a series of these linked together and standing side by side. Within the cabinet are shelves with slots (grooves) along the bottom into which circuit boards for different functions are slid. Some PBX manufacturers make separate modules (like a shelf within a separate housing) which stack on top of each other and may ultimately get up to that refrigerator size as the PBX grows. Some newer PBXs operate using one or a group of PCs to house the circuit boards.

Digital Station Circuit Boards

The typical PBX works with proprietary telephones unique to that PBX manufacturer. These are called digital telephones and come with lots of buttons for accessing extensions and the functions of the telephone system.

Telephones (stations) need station circuit boards within the PBX to make them work. How many telephones do you need? How many will you need to add during the life of the system (approx. 10 years)?

Count the people who need a telephone and other locations such as conference rooms, reception areas, lunchrooms and security areas. Equip the PBX with a sufficient number of digital station circuit boards to support this number of telephones. Most PBXs have 8, 16 or 32 ports per station circuit board (a port is an electronic "place" and you need one per phone). Say you need 155 telephones and you buy a PBX with 16 port station circuit boards. You would equip the PBX with 10 digital station circuit boards (16 x 10 = 160 phones supported) and would have 5 spare ports left over. Now you can add 5 more telephones to the system before you have to buy another station circuit board.

It is always a good idea to plan for 100% growth. Growth in the PBX doesn't only have to do with adding people or telephones. You may add outside lines to work with automated answering systems (voice mail, automated attendant). You may add links between your PBX and computers. All this takes up both physical space in the PBX and capability in the PBX software, so leaving room to grow is important.

When you buy a PBX be sure you will have spare space within the cabinet to add more circuit boards or, if it is a modular system, that you will be able to add more modules.

Companies selling PBXs sometimes use the terms "equipped for" and "wired for" to describe the sizing of the PBX. Equipped means that the circuit boards are in place and wired for means that the slots are there to add circuit boards that you can buy as you need them.

PBXs also have software and memory "space" that goes along with the hardware space. More on that to follow.

Another use of digital station circuit boards is to connect the PBX to the Voice Mail. Voice Mail often connects to the PBX via these ports. If this is the case with your system, this needs to be included in the count of circuit boards.

Analog Station Circuit Boards

Originally, this type of circuit board was added to the PBX to support the older style single line telephones that could only answer one extension and had no buttons for accessing PBX functions. Some organizations still use this type of telephone which typically costs a lot less than the proprietary digital telephone. As with digital circuit boards, there may be 8, 16 or 32 ports per circuit board.

The proliferation of fax machines and computer modems has created a new demand for analog circuit boards. The outside lines (trunks) coming into your PBX can be shared by fax machines and computer modems. An extension from an analog circuit board at each fax machine and computer modem location gives that fax or modem access to the outside world. So another aspect of sizing the PBX is to count the number of faxes and modems requiring analog extensions from the PBX. Note: In many offices, these computer modem lines are being replaced with high capacity circuits to the Internet, shared by all office computers.

Trunk Circuit Boards

There are different types of trunks or outside lines that may connect your PBX to the public switched network (the world outside the office). Note: Each trunk handles a single telephone call at a time. See "trunking" below. Once you have decided on the number and type of trunks you need, the appropriate circuit boards to go along with these trunks are ordered for your PBX. As with station circuit boards, each board has ports for either 8,16 or 32 trunks. A T1 circuit board usually handles only 1 or 2 T1 circuits. Again, plan for growth. Find out the maximum number of trunks and T1s that your PBX can handle.

Trunk circuit boards are generally classified as follows:

▷ **DID boards:** These are specifically designed to handle DID (direct inward dial) trunks which enable callers to reach everyone within your PBX by directly dialing a unique 7 digit telephone number.

▷ **Combination trunk boards:** These support trunks which handle both incoming (to the switchboard) and outgoing (dial 9) calls. Sometimes these are called both way trunks.

▷ **Universal boards:** These support both DID and combination trunks and enhance the flexibility of your system. They also make better use of that valuable "real estate" within your PBX cabinet.

▷ **T1 or PRI circuit board:** The T1 is a circuit which provides the equivalent of 24 incoming and outgoing trunks. The PRI (primary rate interface) is a T1 with some extra capabilities such as delivering the telephone number of the person who is calling. Most large organizations (over 150 phones) are now using T1 or PRI. It is important to know how many of these your PBX can accept.

▷ **Paging trunk circuit board:** Paging systems are undergoing a bit of a renaissance as callers tire of voice mail and people within the office become harder and harder to locate. Access to an overhead paging system through the PBX also requires one or more ports on a trunk circuit board.

▷ **Tie lines:** If your organization has tie lines physically connecting you to other sites, there is also a tie line circuit board needed to support this.

The PBX has other types of circuit boards for processing calls and other functions. This varies from one PBX to another. Find out how your PBX is put together and be sure there is room for everything you need now and what you may need in the future.

PBX System Memory

If you open up your PBX cabinet and see a lot of space to add circuit boards, don't assume that you can just fill them all up. Other aspects to PBX growth are the system memory, software and processing power. This is another area where each PBX manufacturer has developed the product line in a different way. Ask how the memory in your PBX is used and what your options are for getting a system with more processing power (and why you might need it). Each PBX has some system maximums for various functions such as intercom groups, call pick up groups (answering calls from extensions which do not appear on your telephone), speed dial numbers, etc. Find out in advance which system functions you will be using and what the limitations are within that PBX. Ask about the different software releases of your system to be sure you are getting the latest version.

As complex as all of this may sound, the material in this tutorial is an oversimplification. Each manufacturer's PBX and its connections to other systems has some unique aspects to it. It is important to get familiar with these to avoid "nasty surprises."

Scaling Voice Mail

Voice Mail is put together in a manner somewhat like a PBX with circuit boards and associated memory and processing power. For sizing a voice mail system, it is important to anticipate how it will be used and to provide plenty of room for growth and flexibility.

Here are some things to think about for sizing your voice mail system.

How many people will be using it simultaneously? Voice mail systems are sold with a certain number of ports, usually 4 port, 8 port or some additional increment of 4 or 8. The average telephone system with 100 people typically uses an 8 port system and finds it to be adequate. This means that 8 people can be using voice mail at the same time.

In thinking about how many people will be using the voice mail simultaneously consider callers leaving messages and users retrieving voice mail messages.

Some voice mail systems have "dynamic" ports which can be used for Automated Attendant (for sales press 1, for customer service press 2, etc.) as well as voice mail. If you decide to do this, consider how many additional people may need to be simultaneously answered by the Automated Attendant. Each of these callers is using up a port while they are directing their own call to the appropriate extension or department (once the caller reaches the extension, the port is free).

Find out how your Voice Mail links to your PBX. The one that have "digital connections" more smoothly integrate with your PBX than the ones which connect to it via analog station ports. In either case, you need a sufficient number of ports in the voice mail to connect to the PBX. In some cases the voice mail may be one or several circuit boards within the PBX, but all of the above still applies.

Another sizing variable for voice mail is the number of hours of storage. As with PBXs, voice mail system manufacturers use software, processing power and system memory in different ways, so it is important to get intimate with your particular system (or the one you are planning to buy). Things to tell your service provider so that the system will support you include:

- Number of users (don't forget people in the "field" who may not even have a telephone on the PBX but may use the company voice mail)

- Length of messages (do you want callers to be able to leave a message of any length or to give them a limit)

- How long will users messages be stored? (You need to set some rules with your staff so that the voice mail messages are promptly retrieved.)

- Plans for growth in the number of people using the system.

If you are going to be using Unified Messaging, where your voice mail messages can be retrieved, viewed and responded to on your PC connected to your local area network, this must also be considered in sizing the voice mail.

Trunking

Like many aspects of telecommunications planning, determining the number and type of outside lines coming into your telephone system **is** an inexact science. It is one area where a bit of overkill is definitely preferable to planning too precisely.

You can never really know how many people will be calling you at the same time or how many of your staff members will be dialing out. The goal of successful trunking is for callers never to reach a busy signal and for your staff to always be able to access an outside line. A further goal is not to install many more trunks than you need, paying monthly for capacity that you never use.

The quick (dartboard) method of determining trunking is to make a judgment based upon the number of people in your system. For example, if you have 100 people and your office has average telephone use, 25 combination trunks (handle incoming and outgoing calls) will probably be more than enough. Most business people don't believe this, since it appears that hundreds of calls are handled every day (and they may be) but it is not likely that more than 1 in 4 people will be on an incoming or outgoing call at the same time. As the number of people in an organization increases, the 1 in 4 proportion changes. For example, if you have 400 people it is likely that 100 trunks would be excessive.

If you are using both DID (incoming only) and combination trunks (typically used for outgoing only when you use DID for incoming) since you are designating specific trunks for incoming only and other trunks for outgoing only, you will need a greater overall number of trunks.

If you are using a T1 circuit, the same concepts apply since each of the 24 channels of the T1 can carries either a DID trunk or a combination trunk. The newer PRI circuit provides more flexibility in that each of the 23 channels of the PRI can carry either an incoming direct dialed or outgoing call, each channel being like a combination trunk and DID trunk all in one.

There is a discipline called "traffic engineering" that uses statistical probabilities to determine the number of trunks needed. In order for traffic engineering to be useful, you need to have good information about the number and length of incoming and outgoing calls your organization handles in the busiest hour of the busiest day at the busiest time of year. Most organizations do not have this information, but it is obtainable. Your PBX vendor can run a traffic study to determine the number and length of calls, hour by hour over a period of time. If you are going to do this, pick a busy time and have the traffic study run for 1 to 2 weeks. If you have a Call Accounting system in house (keeping track of who made which outgoing call) this can also be set up to run a traffic study. Another way to get some call volume information is from your telephone bills, but unless you use only 800 numbers for incoming calls, the bill information will give you complete outgoing calling volume only.

Once you have your busy hour information, it is best to seek out a true traffic engineering expert to interpret this for you. Your PBX maintenance company may have such a person on staff or you may find one working at a telecommunications consulting company. Traffic engineers use statistical tables that have been developed, such as Poisson and Extended Erlang B, which estimate the number of trunks required to handle a given amount of telephone call volume. These tables require that calling volume be translated into a unit called a CCS which represents one hundred call seconds (100 seconds of conversation) or an Erlang which represent one hour of conversation (36 CCS). There are also some software programs you can purchase that have the statistical formulae built it, but these programs can be dangerous in the wrong hands (garbage in, garbage out) and you would be better off going back to the dart board method.

Provisioning the Surround Stuff

Some callers may prefer
automated systems to real people!

The "Surround Stuff" (a term coined by Harry Newton) refers to all of the information that people call your organization for on a regular basis. The "stuff" is the many details that surround any business, from the trivial-but important ("Where are your offices?") to the vitally important ("It's 7:00 P.M., the stove you sold me has just broken. I need to cook dinner for 11 people. How do I replace the fuse?")

The objective is to automate this surround stuff as much as possible, to provide fast, 24 hour, 7 days a week response, keep customers happy and to not waste expensive staff time on routine issues. Deploying these systems well, in the right order and continuing to improve the way they handle callers is key to growing a manageable, cost-effective enterprise that doesn't alienate its customers.

Identify Your Surround Stuff

Conduct an organization wide effort to identify all types of calls you receive and make. What information is given out? How does your staff get to it? How frequently is the information requested? How is it disseminated (by phone, fax, e-mail, regular mail?) This may be your hours of operation, your fax number, an explanation of how you do business (if all the salespeople are out it the field, it's better than having a clerk take a potential new client call), instructions on how to operate your products, the status of a customer's order, etc. Stretch your imagination, but don't tell your staff you're considering automating these functions, only that you're trying to make the systems in your company more efficient. This avoids creating unrealistic or premature expectations.

Consider Types Of Technology Currently Available

To keep current on what's possible, and to keep up with your competitors keep you subscriptions to telecom and customer service industry publications up to date! Here are some of the systems currently available to you:

▶ Automated Attendant enables a caller to direct his own call to the right person or department. "For sales press 1, for customer service press 2, etc." This can also be used to direct callers to more surround stuff, "To use our automated system to check the status of your order press 3" or "To receive a fax with directions to our store press 4." Automated attendant can be a separate device or a part of your voice mail system.

▶ Voice Mail is primarily thought of as a place for callers to leave messages, but it can be effective for giving out information as well. Experiment with setting up voice mailboxes providing information only or also enabling a caller to leave a message after hearing the information (if you do this, make sure your staff has the time and motivation to return the calls.) Callers can reach these Voice Mail announcements through the Automated Attendant or can be transferred to them by anyone in your organization (make sure everyone knows how to do it.)

▶ Interactive Voice Response (IVR) gives callers access to information that is more complex or that changes regularly. The callers uses the touch-tone buttons on his telephone (or speaks certain words if the system uses Voice Recognition) to request information. When calling your bank to obtain your checkbook balance or to find out which checks have cleared, you are using an IVR system. The database being accessed by the caller may reside within the IVR system or may be on another computer to which the IVR system is connected. If you are in the process of setting up a computer database which you may want callers to access using IVR, this can be a part of your planning process. Keep the menus as simple and straightforward as possible and keep the number of submenus to a minimum. The way you get to the information from a computer screen is emulated by the callers using the IVR.

▷ Interactive Fax Response goes by several names including Fax Back and Fax on Demand. The equipment that supports this is the Fax Server. It works in a similar manner to IVR and indeed may be a subset of an IVR system. The idea is that a caller requests that information be faxed to him. He may be offered a list of documents, for example, or, after getting his bank account status, he may enter a fax number to have a printed statement faxed to him.

▷ A web site on the Internet is yet another way for customers to communicate with you and to obtain information on your company or on their particular account status. The web site can now tie in with your telephone system and office e-mail system enabling people at your web site to send you an e-mail or request a call back.

Connecting Your Surround Stuff To The Outside World

Other than the web site, these systems are typically housed on your premises and must be accessible through outside telephone lines. Most have the ability to connect directly outside lines (some only a T1 type of circuit), but more typically, they are connected through your office telephone system (the PBX). This enables callers to more easily get to a live person if they wish to exit the automated system. Part of provisioning is being sure that your PBX is up to the task. Get your PBX supplier involved early in the process to determine if you will need any PBX upgrade or expansion. The surround stuff may be connected to your PBX using what is called an analog port (the same type of port to which you'd connect a basic single line telephone). It may also connect using a digital port (the same type of port to which you'd connect the manufacturer's proprietary digital telephone) or to a T1 circuit board (similar to what you'd use to connect the PBX to a T1 outside circuit). The number of ports affects the number of people who can be using the automated system at the same time.

Find out what will happen to callers if your automated system is busy. What will they hear? Not dead air! How about a nice promotional announcement followed by offering the caller the option of waiting for a live representative or leaving a voice mail message. What will happen to them after 3 minutes of waiting time? They won't be disconnected! Give them the option of waiting longer or leaving a voice mail message. Making sure that the automated systems cooperate with your PBX and voice mail to handle callers at the busiest times of day is critical to the success of the automated system.

Newer systems may provide the PBX and the surround stuff capability all on one platform. These typically require a significant amount of customized programming and are not yet widely being used, but bear watching.

There are also Service Bureau companies from whom you can rent and experiment with these automated capabilities. The down side to going this route is that it can be more cumbersome to keep the information in the systems current.

▓ Implementation Tips

▷ Start with something simple. Get comfortable with it and measure how your callers react. Then start planning for something more complex.

▷ Put together an "in English" statement of requirements describing exactly how you expect the system to work, including what callers will hear and what will happen to callers under a variety of circumstances.

▷ Look for a supplier who has experience in your industry and with your size company. Most automated systems require a large amount of custom programming. Certain suppliers cater to vertical markets. For example, many IVR systems have already been built for the Employee Benefits area and for the Real Estate industry. Why build an entire system if something is already available which will do the trick or may be slightly modified to meet your objectives?

▶ Look for a supplier in your neighborhood who can support the application (both hardware and software). A significant amount of time, energy and money can go into both the building of these systems and their care and feeding. Many a buyer has spent thousands of dollars on airline tickets and hotel rooms for supplier support people. If you buy a system from someone in another part of the country, closely investigate the local support capability. These third-party support relationships are often short-lived.

▶ While some PBX suppliers have experience with these interactive systems, this is not the rule. If your PBX supplier proposes to sell you an IVR, find out how many of these they have installed and support. With all suppliers, check references and try to visit some other customer sites.

▶ Don't rush the development of your IVR, Fax or Web site system. Plan time for testing and making adjustments. The more complete it is when you finally roll it out, the better it will be accepted by callers (users) and staff (yes- the staff has to be as comfortable as the callers!).

▶ Unless you have real expertise and experience on staff, hire some professional help to specify the system requirements. Most suppliers who sell these systems also sell consulting services, which you will need to develop your applications. You may want your own independent consultant representing you, as well.

▶ Most surround stuff suppliers sell hardware along with their software. While it may make sense to purchase the hardware from them to establish a single point of responsibility, find out who will actually support the hardware. Sometimes companies who view themselves as being in the software business do not see the hardware as part of the "system" and may leave the customer to fend for himself on hardware problems. You may need a separate service contract with a computer support supplier. The hardware will typically be a PC, monitor, printer, and back up system.

▷ Find out what you're actually buying, in terms of the software. If it is a license, how many people can use it and how is this determined.

Administration Tips

▷ The surround stuff consists of both hardware and software. Keep records of what you have purchased and documentation of how it has been set up. Keep up to date with software releases. When you purchase the system, ask the supplier if they will sell you a program to keep your software up to date.

▷ Ask the supplier what they provide in the way of initial documentation and how they keep this up to date as the system grows and changes. If they don't keep it up to date, you will designate someone in house to do this.

▷ Find out what reports are available from the system and use the information to fine-tune the caller experiences. Identify bottlenecks or points at which callers may become confused and exit to a live person. Expand options in areas where caller interest in the highest.

▷ Develop a system for keeping the information in the system database current and updated regularly. You may need a full time or part time staff person for this depending upon the complexity of information being maintained and how often it changes.

▷ As much as is possible, if you are buying multiple "surround stuff" systems, keep with the same platform which will make administration easier.

▷ Test before you implement a system and set up a plan for ongoing testing to be sure all aspects of the application work. Pretend you are a caller and try the system in every imaginable way, including the wrong way. Try to crash it and be sure you can't!

The Finer Points of Managing Telephone Systems

The job we love to hate!

How to Manage Your Telephone System

Don't just set it up and forget it!

The Telephone System Defined

The complete definition of a telephone system includes the lines connecting your organization to the outside world and the cable within your walls. It also includes peripheral systems such as Automated Voice Processing. While these are part of the overall system and things would not work very well without them – the "system" on which we are focusing in this tutorial is the PBX (private branch exchange) or the Communications Server as some newer computer based telephone systems are being called.

Why Manage It?

Introducing the concept of "managing a telephone system" often draws puzzled stares. Everyone can relate to buying a system, setting it up and having it installed, but *managing* it? Here are some manifestations of not managing the telephone system.

Poor Service Levels

▷ Callers complain of waiting a long time to be transferred into voice mail so that they may leave a message.

▷ Callers using the "spell by name" directory to locate someone within your organization do not find the name of someone who works there.

▷ Evening callers hear a recording that they are being "transferred to the attendant", but there is no live attendant (switchboard operator) to answer.

▶ Direct dial telephone numbers within your organization have two different 3 digit prefixes (or even worse, 2 different area codes!)

▪ Lack of Expense Control

▶ You pay $2,500. for a new circuit board to support additional outside lines. When the technician installs it, you learn that you had spare capacity on existing circuit boards and did not need the new one.

▶ When someone wants to add one more telephone, you discover that your system is at maximum capacity unless you invest in another cabinet or server at a cost of $4,500. This expense has not been included in your annual budget.

▶ You discover that the "automatic route selection" in your telephone system has been sending long distance calls over the wrong set of outside lines, resulting in a 25% increase in your long distance billing for the last 5 months.

▶ You find a closet full of disconnected telephones for which you paid $600. each. Since you didn't know you had these, you've been buying new ones!

▪ Inability to Accommodate Changing Business Requirements

▶ Your boss announces that the competition is using "screen pops" enabling customer service representatives to view a screen of client information as the call arrives at the desktop. You call your telephone system vendor and determine that the software upgrade you just purchased will not provide this. You can get the capability with a new software upgrade available in the third quarter of next year. (You'll be the first to try it!)

▶ Your ad agency wants to launch a campaign using 800 numbers in ads that will run this month. They may generate a 50% increase in calls coming into your system. You find that the additional circuit boards you will need to handle this increase are back-ordered and not available till next month.

▶ Your organization is currently at three different addresses in the same city. Each has a telephone system less than one year old. Now all three locations are consolidating into one and you want to reuse the equipment. You discover that it is not compatible and must purchase a whole new system. No parts of the other systems can be reused. Your boss is unhappy!

Managing the telephone system means good record keeping. It means continual monitoring of how the system is working. It means planning ahead and keeping up to date on the products you own and what the latest versions of those products can do for your company. It means keeping close relationships with your vendor representatives.

Following are some ideas to get you started.

Record Keeping

Good record keeping is essential to telephone system management. It's amazing how few organizations do it well. These are the records to keep:

Record of All Outside Lines Connected to Your System.

▶ Type of line (combination trunk, DID trunk, T1, etc.)

▶ Telephone number or circuit number associated with the line

▶ Date of installation (and disconnect if removed)

▶ Order numbers for installation and removal.

▶ Contact names, addresses and telephone numbers company providing you with the line

▶ Location on the demarc (main distribution frame from local telephone company) in your telephone room (if the line goes through more than one demarc, it's a good idea to know where on each demarc – helps with trouble shooting. Don't expect your telephone system vendor to keep this information. They will charge you for their time looking around if the demarc is not clearly labeled and associated with each outside line.)

▶ Location (circuit board and port) in the telephone system.

▶ Monthly cost of the line (to be checked regularly against your billing detail)

▶ Any other characteristics of the line – depends on type of line (for example – on a combination trunk, is it loop start or ground start? – On a T1 is it ESF or B8ZS signaling?)

▶ Any other hardware external to the telephone system associated with the line (channel bank, multiplexer, etc.)

● Record of Your Telephone System Circuit Boards and Ports

▶ Type of circuit board (e.g. trunk, station, T1, tie line, processor, special function)

▶ Number of ports on each circuit board.

▶ Shelf and slot number location for each circuit board.

▶ Associated outside line number or station number (extension number) on each port

▶ Spare slots on each shelf and what type of circuit board can go into that slot.

▶ Cost of each type of circuit board

▶ System expansion capability within the existing cabinet and with added cabinets (or with communications servers.)

● Record of Your Telephone Instruments (Stations)

▶ Type of instrument (Name, model number, number of fixed function and flexible buttons)

▶ "Face layout" meaning which extension numbers appear on which buttons and which system features appear on other buttons. (Find out if your telephone system can print this out for you before doing a manual inventory – the labels on a system that has been installed for awhile are often incorrect. Accurate labeling of telephones is part of good telephone system management. Labels should be typed, not handwritten.)

▶ Associated shelf, circuit board and port in the telephone system cabinet

▶ Location number on a floor plan.

▶ Associated cable number (each end of each pair of wires should be labeled both where the telephone plugs in at the desk and back in the telephone equipment room where the system control cabinet is located.)

▶ System programming associated with the telephone which cannot be determined by looking at the telephone (e.g. call forwarding when the telephone is unanswered or busy and toll restriction – meaning which areas can be dialed from that telephone, etc.)

▶ Stored inventory of unused telephones and circuit boards (a good idea to keep some spares on hand.)

● Record of Used and Unused Extension Numbers and DID Telephone Numbers

▶ EXTENSION NUMBERS

The programming of your telephone system may use digits for some system functions. For example "dial 9" used to access outside lines. When you are setting up the system, try to keep as many digits free as possible for use with system extensions. For internal dialing it is cumbersome to have extension numbers of more than three digits, unless the size of your system forces this. So, for example, your extension groups may be 200-299, 300-399, 400-499, 500-599 etc. Find out which groups you have available and keep track of the use of these extensions. Make sure you have enough to accommodate the anticipated growth of the system. If possible, when setting up the system, you can assign like extension numbers to specific departments or floors. This helps with record keeping, but tends to deteriorate over time if people in your organization move around a lot and need to take the extension number with them when they move.

If you have not kept track of extension numbers, your telephone system vendor can help you to figure out which ones are in use and which ones are spare. This should be accompanied by a physical inventory, since an extension can appear to be live in the system cabinet (associated with a station port) but there may be no telephone at the desktop end of the cable (some systems will detect this, but not all).

Some extension numbers are considered to be the primary number associated with a specific telephone instrument and other extensions may be "virtual" meaning that they appear on a button of the telephone, but are not the primary extension . Check with your telephone system maintenance company on use of extension numbers and how to manage them.

▶ DID (DIRECT INWARD DIAL) EXTENSION NUMBERS

Everything we've said above about extension numbers also applies to DID extension numbers. The difference is that, with a regular extension number, callers cannot reach your desk by dialing directly to a 7-digit telephone number that is uniquely yours. Within the telephone system the DID extension may be either 3 or 4-digits (largely a matter of preference). For example, it your DID telephone number is 628-7422, your internal extension can be either 7422 or 422.

While regular extensions are part of your telephone system only, DID extensions correspond to DID telephone numbers purchased on a monthly basis in blocks of 20 or 100 from your local telephone company. For example, your organization may have reserved the following blocks of numbers 628-7400 through 7499, 7500 through 5599 and 7600 through 7699. If you think you will need more DID numbers over the life of your telephone system reserve them up front. Otherwise you may not be able to get more blocks of numbers with the same 3-digit prefix. Having 2 different prefixes can get very confusing for people calling your company.

Another thing to remember, is that if you are assigning DID extensions, people who need more than one extension to be able to handle multiple calls only need one DID number assigned to them. When the DID extension is busy calls roll over to a non-DID extension.

● Telephone Directories

Directories keeping track of how to reach people are a great management challenge. There are telephone numbers, fax numbers, cell phone numbers, e-mail addresses, etc. and all must be continually kept current.

When managing the telephone system, it is important to keep the number of separate directories to a minimum and to figure out how they will all be kept up to date and in sync. For example, some PBXs have a directory built in associating a name with each extension. If you have a separate voice mail system, that also has a directory with names and mailboxes. A Call Accounting/ Chargeback system tracking who made or received which telephone call also has a directory. Then there may be a separate directory for your organization's computer network and e-mail addresses.

Other Telephone System Programming

Keep track of all telephone system programming and allocation of memory.

▶ Automatic Route Selection (also called Least Cost Routing) programs your telephone system to send calls out over the lowest cost group of outside lines. You may have a dedicated T1 circuit to your long distance company, for example, which enables you to make calls at a lower cost per minute that over regular trunks. It is important to work closely with your system vendor to program the ARS (the telephone system vendor does not know what arrangements you have made for billing with your local and long distance telephone companies.)

▶ Find out what the capacities are for all aspects of system memory and how the memory is allocated. In some areas you may have a choice here and in others you will not. The more you understand about how your telephone system is put together (they're all different) from a hardware and software perspective, the better you will be at managing it.

Telephone Bills, Purchase and Maintenance Contracts, Leases

While Managing Telecommunications Expenses is a whole different area than managing the phone system, the two are closely related. Trying to manage the telephone system without an awareness of the bills, contracts and leases is not doing the complete job. Often, in larger organizations, the people who manage the telephone system never see the bills, which go directly to accounting. The accounting department pays them without really knowing what they're paying for.

Here's what needs to be looked at, understood and managed.

▶ TELEPHONE BILL RECORDS

▶ A copy of all pages of each bill for telephone lines, usage, equipment and services received on a monthly basis. (Some local and long distance companies provide on disk or CD-ROM. We suggest getting the main pages of each bill on paper as well, reserving the call detail for the disks. Some people never get around to looking at the disk, but the ability to "eyeball" a paper bill can uncover any glaring errors or increases.)

▶ A copy of billing detail (often called the Customer Service Record by the local telephone company) which lists all outside lines and services and the associated monthly charges.

▶ PURCHASE CONTRACTS

▶ Keep copies of the purchase contract for the original system and all expansions, additions and upgrades. If you don't you may wind up paying for the same thing twice because you don't know that you already bought it. Don't rely on your vendors to keep this information. This includes keeping contracts for local and long distance telephone service. (You may be paying more than you agreed to.)

▶ MAINTENANCE CONTRACTS

▶ The telephone system maintenance contract typically covers all repair calls, labor and parts for anything that goes wrong with the system. It may be paid for on a quarterly or annual basis and must be adjusted upward or downward as your system expands or contracts. It is most often based upon some amount per month times the number of working system ports.

▶ Keep track of when your maintenance contract is coming up for renewal. You may want to get competitive bids if you're not happy, although keeping the vendor who installed the system and is familiar with it is generally a good idea.

▶ You may also negotiate "value added" items such as an annual traffic study to monitor use on your outside lines into the maintenance contract.

▶ Monitor ongoing agreements you may have with local and long distance companies, too. Don't be shy about asking to renegotiate before the end of the contract if you think you're paying too much since rates are dropping.

▶ LEASES

▶ Keep track of all leases in effect related to the telephone system and associated equipment and services. When were they initiated? What did they cover? What cost was financed and what are the terms? Is it a $1.00 buyout lease or a Fair Market Value buyout? If the latter, how will the fair market value be computed.

▶ Start looking at your leases well before they're up so you will have sufficient time to make a good decision, such as whether to continue to lease, buy out the lease or give back the leased equipment (you can be sure the leasing company does not want you to give it back.) Beware of leases that "roll over" automatically if you miss the lease renewal date.

Work Order Tracking

Good record keeping in all of the areas covered above requires that you track any changes you make to the system. Even something as simple as adding one extension number to a telephone or moving a phone from one office to another requires that all associated records be updated.

There are software programs designed to manage all of the above information and to track work orders. We suggest doing it manually first unless your system is very large. Once you have accurate information and know what you need, you will be better able to define how you want to manage this using software.

Ongoing Monitoring

While good record keeping is essential, it does not guarantee that your telephone system actually works well. For this, continual monitoring is needed. Most telephone systems are not completely tested when they are installed. Problems with system functions arise as people encounter them and complain about them. Some system problems are never fixed. This reflects poorly on your organization. Take the time to do it right Here are a few suggestions.

Call Into Your Office

▷ Call into your telephone system and see what you encounter. Call at different times of day and at night when your organization is closed. How long does it take to answer? Do you get to the right person on the first try? If you ask to be transferred, did it work?

▷ Ask to be sent to someone's voice mail? Does this work? How long do you have to wait? Do you hear an appropriate greeting? Can you easily escape from voice mail to a live person and how does the system instruct callers to do this?

▶ How does your Automated Attendant work? Is there a company directory? Does the "spell by name" work easily? Try to "crash" your automated systems by dialing random digits as someone unfamiliar with your system may. Try doing nothing and see how long you wait before a live person answers your call.

Test System Functions within the Office

Make a list of all system functions and test them periodically or build regular testing into your maintenance contract. Call Forwarding when extensions are busy or unanswered often goes awry. Things like being able to transfer a call or set up a conference call seldom fail, but what does fail is people's ability to use these functions. Regular *training* is also a big part of telephone system management.

Traffic Studies

A traffic study monitors the volume (total calls and total length of time) of incoming and outgoing calls on your outside lines for a given time period. Most traffic studies run from 1 to 2 weeks. The purpose of this is to ensure that you have a sufficient number of lines connecting you to the outside world (local and long distance companies) to handle your call volume. The traffic study can identify outside lines with no usage which may indicate that the line is not working (or a bad port on the telephone system circuit board). It can also identify when you have many more lines than you need.

An annual traffic study (including an interpretation – not just raw data) can be negotiated with your telephone system supplier as a part of the annual maintenance contract. Otherwise they may charge for the study.

You can also set yourself up to do traffic studies in house with a Call Accounting system.

● Testing of Automated Voice Systems

There are services used to test Automated Voice Processing Systems, simulating real incoming telephone calls and identifying bottlenecks, inappropriate recordings and other problems your callers may encounter which translate to "poor customer service."

● Remote Monitoring by your Equipment Vendor

Most telephone systems are sold with remote monitoring capabilities. Some systems will dial out to the equipment vendor to indicate a problem. Other vendors may monitor systems regularly. This is good preventive maintenance. Find out exactly what your vendor is capable of, what is being monitored and what the schedule it. It is still a good idea to schedule and pay for (or write into your maintenance agreement) an annual on site check up by a system technician. This can also help you to reconcile any record keeping discrepancies that may have developed.

■ Planning Ahead

● Expansion

When you purchase the system, and at regular intervals in the life of the system, find out what growth capacity you have both in terms of system hardware and software. Determine what different expansion increments will cost before you need them. System expansion is needed even if you do not plan to add more people. If you plan to use automated systems your telephone system needs to expand both for the interfaces to those systems and for the additional outside lines handling calls to the automated systems.

● Software Upgrades

Keep up to date on the latest software upgrades from the manufacturer of your telephone system and on what changes are included in the upgrades. Whether or not you need the upgrade for added functions, it is a good idea to keep your system at the latest software release. Older software releases may develop flexibility or service problems over time as fewer people are available to support them. Also, upgrading gradually can cost less than moving up several levels all at once.

We recommend discussing software upgrades when you are purchasing the system. That's when you have the most leverage. What software upgrades have taken place since the system first came on the market? What has been gained (or fixed) with each subsequent release?

Business Partners

All telephone system vendors have realized that they can't provide everything their customers may want. As telephone systems merge with computer systems in varying forms of integration, manufacturers and suppliers are forming relationships with other companies to provide complimentary products and services.

Dig beneath the surface to find out the true nature of these partnerships. How many customers have used the product of the business partner with the telephone system and what has the outcome been? How long has the relationship lasted (many tend to be fleeting.)

Keeping Good Supplier Relationships

Salespeople and Technicians

Telecommunications system suppliers have many customers clamoring for attention and decreasing resources with which to respond. Take the initiative to set up regular meetings with both a salesperson and the technician who regularly services your system.

The person who sold you the system may not be the "installed base" salesperson who will service you on an ongoing basis. This person should bring you up to date on what's new with the company and the telephone system.

The technician familiar with your system will likely provide a more down-to-earth view than the salesperson, and may alert you to problems you may not have otherwise discovered. It is definitely a good idea to have one or two regular techs servicing your account, who can get to know your system and how you use it.

Be a Good Customer

Have a heart to heart talk with your supplier representatives about what support they can realistically provide and what you need to do in-house. Many organizations tend to ignore the principles of good telephone system management and then "blame it on the supplier" when things go wrong. This does not endear you to your supplier and may result in their focusing their energies where they are more appreciated.

As the Communications Server begins to replace the more traditional system, the need to manage your systems in house will increase. Coordination among multiple suppliers will be the order of the day and System Integrators will replace the more conventional telephone system supplier and maintenance company.

A Window On Your Business...
Using The Telephone System For
Management Reporting

You never know what you'll find!

Introduction

Managers have historically viewed the telephone system to be in the same category as the typewriters and paper clips; an uninteresting but necessary expense. Now, as these same managers read the morning paper, the word "telecommunications" is everywhere! The message from the media is that the way the organization applies telecommunications technology is changing. Upon entering their office, the managers encounter a telephone system that does not reflect any of this excitement they've just read about! How will these two different realities be reconciled?

If you're blessed with the responsibility for your company's telephone system, great opportunity awaits you. Make a plan, start small and begin to change the way management views corporate telecommunications. Here are some ideas to get you started:

Find New Ways To Use The Information You Already Have

The Long Distance Bill

We'll grant you that long distance bills can seem unfathomable, but with a bit of attention to figuring them out, they can contain a wealth of information about the activities of your organization.

1. What is the average cost per employee for long distance calling in your organization? How does this compare to other

organizations in your industry or of your size? Is there a trend for cost per employee (up or down) or is there some seasonal variation?

2. Where are the long distance calls going? What does it say about your company's business if in June, 60% of all calls went to Los Angeles and in December, 60% went to Dallas and only 10% went to Los Angeles. The changing patterns of telephone calling can provide management with useful insight as to what is going on in the business. If it is not clear why the calling patterns are changing, it may also provoke questions about what really is going on. Either way, it will get their attention.

3. Where are calls to your organization originating? You can get this if you have an 800 (toll free) number or if your telephone system is set up to capture the calling number of the incoming calls. Does new business come in over the telephone? Management will have an interest in where the action is. For example, if your organization just opened up distribution in the Midwest, is this reflected in more calls coming in from these states? If there was an ad campaign focused on the state of Florida, did the call volume from Florida reflect this?

The Call Accounting System

Many organizations have call accounting systems spewing out reams of paper reports which no one ever looks at. Some smarter organizations are reviewing the way call accounting information is captured and distributed, recognizing the information about the organization that this can provide.

1. How do call volumes to your company vary during the day, week, month and year? How much of this is for incoming calls vs. outgoing? This can be of interest to management to support assumptions about staffing requirements. It also provides another insight into the activities of the organization.

Is the average number of calls per day 2,500 in January and 1,000 in June? This can also be matched up to the origin or destination of calls to provide even more enlightenment. Are call volumes heaviest between 12 and 2? Maybe "lunchtime" needs redefinition.

2. Take a look at the internal calls made within your office? Who is calling whom, and what does this mean? If the art department is on the telephone all day with production, maybe these two departments should be co-located? This type of information helps managers to take a fresh look at how the organization works. Good managers know that is an ongoing process.

3. Are you planning to capture the originating number for incoming calls and cross reference it with your customer database? The call accounting system can be set up to capture the calling number to see how many of your calls carry this information and how useful it is, before you embark upon the more costly process of linking the calling number to a database.

4. Part of being competitive is being able to deliver your product or service at a lower price than others in your industry. Management needs to see how much it is costing to run the organization. The more managers can "drill down" into the organization, the better they can see what is really going on. Call Accounting Systems (or Call Accounting Service Bureaus) not only provide detail on long distance costs, but can support the process of charging back all telecommunications expenses to each individual or department. This may include outside lines, local calls, telephones system purchase and maintenance, cable, PCs, computer network infrastructure and use of internal help desk services. This supports the trend toward making each division of a company accountable for it's own profitability. Managers can also see which departments are spending more per person than others and begin to determine why.

5. Instead of distributing paper reports, distill the information into useful summaries which can be distributed via e-mail. Make it easy and quick for the manager to see and digest the information.

Edge Into Converging Telephones and Computers

Managers are more interested in the computer system than in the telephone system. This is probably because they get more useful information about the business from the computers. How much did we sell last month? How much did we spend? Most managers are responsible for making an organization more profitable, so this is the focus, as it should be. If the information can be made even more useful by combining information about telephone calls with computer reports, this will get management's attention. Here are a few examples:

1. Managers walk through their company and see lots of people on the telephone, but have no idea whether this activity is resulting in the organization making any money. One thing converging computers and telephones enables is the linking of telephone calls to transactions. Of 100 telephone calls in the first hour of the day, 40 resulted in a sale, 20 were complaints, 20 were changes to existing orders and 20 were cancellations. The cancellations calls averaged 10 minutes and the sales calls averaged 5 minutes. Employees with the organization 10 years or more take an average of 2 minutes to complete the sale, while new hires take an average of 11 minutes. This type of information can be invaluable to a good manager.

2. Improve your customer service by linking your company database to incoming telephone calls. When a customer calls, you can identify them by either the number from which they are calling or by asking them to speak their name or enter a customer number. As their call arrives, a screen of information about them (past orders, service history, etc.) can arrive at the

desktop along with a call. This is called a "Screen Pop." It can be very powerful in terms of enabling you to have more enlightened conversations with your callers, providing more personalized service and shortening the time required for telephone transactions.

Create reports that show customer's patterns for calling your company. Our biggest customer calls 10 times per week, usually between 10 and 11 A.M.

This customer waits an average of 5 minutes on the line to speak to a representative.

To address this in a report you may recommend getting a new telephone system which will help you to prioritize incoming calls to connect the best customers to the representatives first.

3. Look into repetitive processes which can be automated, such as callers asking for account balances or employees wanting information about their benefits. IVR (Interactive Voice Response) technology is routinely being used and can free up costly staff for other tasks. Efforts such as this can really get management's attention.

   You may report that currently, 2 people are kept busy 5 days per week during business hours giving callers information on the status of their orders. This costs your company $125,000. per year. For a $75,000. investment, an IVR will give out the same information 24 hours per day, 7 days per week. In addition, it will compile information about the number of calls and what information was provided; for example, how many callers were told the order has been shipped, how many told it will ship within 5 days and how many told the items were backordered?

4. Determine how your organization is using the Internet. Can faxes currently sent over the regular telephone lines be sent through the Internet? Can many of your international calls be replaced with e-mail? This is another area that management is reading about. Most managers appreciate creative thinking in

terms of how these new technologies can be applied to improve operations or lower costs.

■ Communicating Effectively With Management

"Management" is a broad term. Find out the responsibilities of the manager to whom you are reporting and tailor your comments accordingly.

Here are some things that are on most manager's minds:

▶ Managing the Profit & Loss of the company, division or department. If the manager is judged by how well his area of responsibility is doing in terms of generating income and keeping costs down, this will almost always be at the forefront of the his consciousness.

▶ Keeping employees productively working on the activities which most directly affect the Profit & Loss.

▶ Keeping the manager's boss or the company board of directors and stockholders happy with the manager's performance.

▶ Keeping the customers of the organization happy, recognizing that they are creating the revenue .

▶ Having a rewarding life outside the organization.

It helps to personally know the managers to whom you are making reports. Since each person is an individual with a different preference, it is advisable to customize the method and format of your communication for the individual. If more than one person is receiving it, try for a consensus!

Telecommunications Housekeeping Tips For Any Season

Get rid of the cobwebs!

Glance back, look ahead — here are some telecommunications housekeeping suggestions to get you off and running for a new year!

Directory Listings

Many people reach your organization by looking you up in the telephone book or calling directory assistance. Check both the book (white and yellow pages) and directory assistance to be sure they are giving out the correct number and that your listing reads the way you want it to.

Think about listing multiple telephone numbers, under your main listing, enabling callers to reach the correct department on the first try instead of dialing in and then being transferred.

If you have Yellow Pages advertising, review your contracts, consider alternatives and ask your marketing department if they are tracking the business generated from this source to see if its worth the investment.

Direct Dial Telephone Numbers

If you use direct dial telephone numbers, make a list of all number ranges (e.g. 212-883-1000 through 212-883-1299) Make sure you know who is using every number and validate this by having someone dial in on each number. Identify all unused numbers and be sure they are available when you need them for a new person or department. Commit to keeping the records for this up to date.

Always keep a spare block of direct dial numbers (usually sold in blocks of 20 or 100). The monthly cost is low. With telephone numbers running out, this becomes even more important. It would be unfortunate if different departments within your organization had different area codes! This can happen.

If you used DNIS (direct dial type of 800 numbers), the same suggestion applies.

Long Distance Billing

Look at the monthly cost over the past year. Has it gone up, down or remained about the same? Set up a simple spreadsheet to look back at this and keep the spreadsheet up during the coming year. Track the total billing and then the billing in categories such as intra-state, inter-state, international, fax calls, modem calls, outbound vs. inbound 800, etc. Don't make it too complicated or it won't get done. This can be invaluable in helping you to keep control and know how your doing.

Meet with your long distance carrier representative. Determine if the rates you are paying are competitive with what similar sized organizations are charged. Make sure you have a fixed cost per minute for each category of call, making it easier to track. If not, ask to renegotiate rates even if you are still within a contract. They'll do it. Set up quarterly or semi-annual meetings to review billing.

Ask your long distance carrier to validate past billing for accuracy. If it is not clear that the billing is accurate, think about hiring an outside firm to review past billing. This can be done on a contingency basis so you only pay if any money is recovered.

Take similar steps with your local telephone service provider.

Traffic Study

Pick a busy time during the next few months and arrange for your telephone system maintenance company to run a traffic study (or run one in house if you have the capability) Include all outside lines including T1s. The traffic study will let you know if you have enough outside lines to handle the volume of calling into and out of your organization. If you have too much capacity, you can remove some outside lines to lower costs. If the study shows that all or most of your outside lines are busy at certain times, start planning to add more before your callers start to reach busy signals. Adding outside lines requires additional ports within your telephone system, so be sure that they are available or added.

Directory Assistance Costs

Look at the cost of directory assistance from your local and long distance telephone bills. If it is significant, consider subscribing to a directory assistance service where the information is provided on CD-ROM and can be distributed on your organization's computer network. Add directory assistance as a category on your expense tracking spreadsheet.

Telephone System Maintenance

Ask your telephone system maintenance company to validate the accuracy of the maintenance agreement. The cost is based upon the configuration of your telephone system, usually by the number of active ports. Most maintenance companies do not change the maintenance cost every time a telephone or outside line is added or deleted, but a once a year reconciliation is a good idea.

If your contract is up for renewal, negotiate in a traffic study as part of the annual maintenance. Also ask about the cost for keeping the telephone system software up to date. If your system has an old software release, you may be missing out on system capabilities and may eventually have support problems. Some maintenance companies will bundle the software upgrade cost into the maintenance agreement.

Telephone Equipment Room

Schedule a cleaning and checking of your telephone equipment room. This room is often treated no better than the mop closet, yet houses thousands of dollars worth of telecommunications equipment. Clean the floor. Look into boxes. Are there telephone sets waiting to be returned for repair.

Check the demarc (cable distribution frame where outside lines are brought in and terminated.) Are all outside lines clearly labeled. Ask your maintenance vendor to check this and to verify that all outside lines are working.

Look at everything that is mounted on the wall. Is there old equipment that is no longer used. Arrange (very carefully!) to have it removed. There's always a danger of disrupting working service when trying to remove old equipment and cabling that is located nearby.

Check record books and repair logs to be sure they are being kept up to date by your maintenance company.

Telephone Instruments

Take an inventory of your telephone instruments. Verify that labeling on all telephones is accurate, not only for the extensions picked up, but for telephone system functions.

Make sure all labels are accurate and printed, not handwritten.

Update your records. This will save a lot of time during the year when changes need to be made. Determine how the records will be kept up to date.

User Training

Convince management that once a year telephone system training is a good investment of time and money. There's no question that it can improve the level of service provided by your organization. Review instruction manuals and have them available during the training. Enlist the support of your telephone system provider, but meet with them before the training sessions to review the plan.

Company Telephone Directory

The new year is a good time to review and update your organization's telephone directory if this has not been a regular activity during the year. Take a look at some of the new facilities management software packages, which will automatically update directory information as telephone changes are made.

Old Telephone Bills And Contracts

Don't throw away old telephone bills. They can be valuable if you decide to audit past billing for accuracy. Your local and long distance telephone companies do not easily provide past billing records. If you get billing detain on CD-ROM save this, too.

Put old bills in boxes, label them with account numbers and dates and store them in archives.

Find all current contracts with equipment and service providers and keep them available, both for validating billing and as a reference point when new contracts are negotiated.

Building Facilities

Call your local telephone service provider to inquire about the capacity of facilities into your building, should you wish to add outside lines (for voice, data, Internet access, etc.) during the coming year. Don't assume that there is a lot of spare capacity available just waiting for you to use it.

This capacity can be pairs of wires running from the basement to your office or can be the cables (copper or fiber) coming into your building from the street. Depending upon the situation, adding this capacity can take several months or more, so don't wait until you're ready to order new lines to investigate this.

■ Convergence of Computers and Telephones

Start learning by doing. Convince management that the organization needs to stay current and begin to add computer intelligence to telephone calls. Find some small application within your organization to begin to experiment. Provide callers with order status information using an Interactive Voice Response system tied to your order processing database. Provide salespeople with contact software that pops up the record of a caller as the call arrives at the salesperson's desk.

Happy Housekeeping!

Training Your Staff To Use
The Telephone System

"If I cut you off while I'm trying to transfer you, please call back."

Why Do We Need To Train Telephone Users?

It's a funny thing. I have been in the telecommunications industry for many years and the focus of my work has been on business telephone systems. Yet when I walk into the average office, I will not be able make a telephone call without asking a few questions. What's going on here?

The reason that end user training is required is that nothing about the telephone system is intuitive. You really do have to be trained and read the manual. Most of the things you are trained on you will never need to do. Most of the things you may want to know are seldom mentioned in the training sessions.

No two manufacturers systems operate the same. There is also no consistency in the naming of system functions. No two systems from the same manufacturer are set up in exactly the same way so there is no such thing as learning how to use a particular telephone. You must also know how that system has been programmed to work.

In this age of increasing automation, the human voice coming from an enlightened person is becoming a powerful business weapon. If you've got good people and want to keep them talking to your customers, it's important that they understand how to operate the tool that lets them do this: The Telephone!

The telephone can work with a conventional PBX or with a PC based communications server. Even if the PC itself is being used as the telephone, this tutorial still applies.

When Do You Need To Train People How To Use The Telephone System?

The purchase and installation of a new telephone system is heralded by a flurry of end user training. This is good and it is needed, but unfortunately, the training typically ends here.

We recommend holding training sessions at least once a year. People forget what they learned initially. As your staff begins to work with a new telephone system, new questions arise.

It is also important to have a training session for all new members of your organization as soon as they start.

If you upgrade your system to add new capabilities or add a system such as voice mail, new training sessions are in order.

If you keep several staff members well versed in the operation of the telephone system, they can be a resource to staff in between regular training sessions.

How To Set Up A Meaningful End User Training Program When You Install A New System

Before purchasing a system, get a very clear picture of the training the vendor will offer as a part of the system purchase. A training class of no more than 10-15 people can be held in just before the new system goes live. It's not a good idea to schedule it too far in advance. It's important for vendor representatives to be on site for at least one week after the installation to walk around providing one-on-one training and answering questions as needed.

As you are purchasing a system, get vendor pricing for training after the system is installed. You can consider packaging an annual training session in with your maintenance contract.

▶ Hold the training at your own premises, in a conference room with working telephones that are set up to work in *exactly* the way the telephones of the people being trained are to be set up. It takes extra effort to have the telephones connected to working lines, but it goes a long way to improve the training experience. Make sure each person gets "Hands on" training at the session.

▷ Make sure the vendor will gear the training to your organization and to how your particular telephone system is programmed. Training is too often generic and not personalized.

▷ Listen to the entire training presentation and review the materials to be sure they are relevant and easy to understand. This is time consuming, but avoids disappointment. The first impression your staff receives of the new telephone system tends to be lasting, so you want it to be favorable.

▷ If you are replacing an old system with a new one, it is useful to have the trainer familiar with the old system operation. That way, the session can relate "Here's how we do it now and here's how we will do it with the new system." This may be for transferring a call, setting up a conference, programming speed dial buttons or any of the features.

▷ Training is not only about which buttons to press to use the system functions. It is equally important for all system users to understand the thinking behind how the system was set up to cover calls. For example, how many times will the telephone ring before forwarding to voice mail? What happens to callers exiting voice mail by dialing "O" – do they go to a switchboard attendant or back to your department? If your organization is changing from a single main telephone number going to a switchboard attendant to a system where everyone has a direct dial number, how will this affect the callers? If the call coverage patterns are different from the system that is being replaced, it is important to point this out.

▷ Train staff on how to answer the telephone. If someone answers their direct dial telephone number with their name only, "John Jones speaking" the caller will have no way of knowing if they have reached the correct company. "ABC Company, John Jones speaking" is better. The important thing is to develop an organization wide policy and let it be known so that callers will receive a consistent and favorable impression of your ability to handle telephone calls.

▶ Having telephones labeled clearly with "in English" easily understood terms for the system functions helps to facilitate training and always makes the telephones easier to use. What does MW mean? Why not label it 'voice mail message waiting' or just 'voice message'?

▶ Keep the training sessions to ½ hour and leave another ½ hour for questions and answers. It is more effective to have several training sessions rather than one that tries to cram in too much information.

▶ Try to make the training entertaining. Use it as an opportunity to get some departmental camaraderie going. Hand something out with the training materials; lollipops, balloons, coffee mugs – kind of corny, but it promotes more enthusiasm.

▶ The department head and management representatives participation in the training process supports the impression that this is important to the company.

▪ Voice Mail

You may want to consider separating the training for Voice Mail from the telephone training. As Voice Mail systems offer more capabilities, the training becomes more complex. How do you set up voice mail distribution lists? How do you receive faxes stored by the voice mail system? As with the operation of the telephones, it is important to establish organization wide procedures as to how these capabilities will be handled.

Training Materials

Training materials prepared by PBX and Voice Mail vendors tend to cover every possible system feature and are therefore not tailored to the way your system is set up and how you plan to use it.

▷ Put together training materials that explain exactly how your telephone system and voice mail systems are set up and how they are to be used within your organization.

▷ Instead of just explaining how to use the system feature, state under what circumstances it will be used and the company procedure. For example, setting up a voice mail greeting should be a matter of company policy. It presents an opportunity for your company to distinguish itself with informative voice mail greetings. The canned "I'm either on the phone or away from my desk" is getting stale. It's much better to say "I'm at a meeting and will return at 3 PM" (as long as you are willing to keep these messages current). The organization needs to take the initiative to ask employees to do this. The point is that good training encompasses a company procedure for how callers are to be handled and for what they hear.

▷ Make up a "quick reference guide" that stays near the telephone. We've put some together using those plastic stand ups that some restaurants use for wine lists. They work very well and do not get lost or mangled as a book or sheet of paper will.

▷ Some telephone systems provide prompts to use system functions in the display on the telephone, which sounds useful, but we do not see this being used.

Speak Up To The System Manufacturer

Telephone systems have evolved over time and some of the operations can be clunky. Some PC based telephone systems replicate this rather than improving upon it. Let your system manufacturer know what you like and dislike about the system. Telephones are often designed with minimal feedback from real users.

Can Your Customers Find You? Keeping On Top Of Your Directory Assistance Listings

"What do you mean our only number listed in Directory Assistance is our fax number?"

Introduction

Do you know how many of your existing and potential customers who need to reach you try to get your telephone number by dialing 411 or the area code + 555-1212? Probably more than you realize. This includes people on the road, people not carrying a totally up-to-date telephone number database in their Palm Pilots, people who lost your business card, and people who just want the convenience of not having to look anything up! These are the people who frequently use what is called Directory Assistance (and used to be called "Information").

Where are these people calling from? It may be their home, their office, a hotel, a pay telephone, a cell phone or someone else's office? For each of these they may be reaching a different company who uses a different database to provide Directory Assistance services.

Are your current and potential customers finding you? And what is the cost if they're not? If someone cannot locate your company in directory assistance, this conveys an impression of an unprofessional business and ultimately translates to frustrated customers and lost opportunities.

Getting It Right The First Time

When you place an order for new business telephone service with a local telephone company, one of the items on the order will be your business listing as it is to appear in the printed telephone directory and in the directory assistance database. Typically, you are entitled to one listing in the White Pages (alphabetical listing) and one in the Yellow Pages (under a heading that groups organizations in the same type of business).

Make sure you confirm your listings in writing (along with the rest of your order). Fax your letter to the specific person at telephone company business office who took the order and keep a copy in your files.

Keep in mind that your billing name can be different from your listing. For example, if your corporate name is Bayville Enterprises, Ltd. but you do business as The Friendly Corner Deli, then The Friendly Corner Deli will likely be the name you want in directory assistance as the main listing.

Additional Listings For Different Business Names Or Different Forms Of The Same Name

For a small amount of money, usually around $2.00 per month, you can request an additional listing with the same telephone number. This is a good idea, since people calling directory assistance may not know the complete name of your business or may know it by a different name. In the case above, the main listing can be The Friendly Corner Deli listed under "T", but additional listings under "F' Friendly Corner Deli, "B" for Bayville Enterprises, Inc. (the corporate name) and another one under "F" for Frank's Deli (another name that people use to identify the business.)

Additional Listings For Individual Names

Some businesses are better known by the name of an individual. So in addition to "Technology Investor Magazine" you may want to list the owner's name, "Harry Newton" (under "N").

Some large organizations, particularly professional services such as law firms, may wish to list the names of partners. If each individual has his own separate telephone number that can be dialed directly, this number can be listed, rather than the main telephone number next to the person's alphabetical listing. Different local telephone companies may have different rules about which numbers can be listed, so check with your own local service provider.

Indented Listings Under The Main Listing

Another option for listing individual telephone numbers by name is to request that these be indented under your main listing. For example, if a caller requests the listing for Hadsell, Overlees & Coe Accountants, the directory assistance operator can ask the caller if he has an individual name at the company. This can save considerable wear and tear on the switchboard attendant answering the main number.

Company departments with separate telephone numbers can also be indented under the main listing. Once the caller gives the directory assistance operator the name of the company, he can then be offered the options of Sales, Accounting, Mail Room, etc.

Directory Advertising

If you want something other than the main and additional listings to make your company stand out in the printed versions of directory assistance (the "telephone books") it is likely that you will be referred to another department within the local telephone company. Directory advertising may be a simple bold listing in the white or yellow pages or it can be as elaborate as a full page ad. (Note: There are competitors to the yellow pages affiliated with the local telephone company, often using a similar name such as the Yellow Book) All directory advertising bears a monthly charge that may appear on your telephone bill or a separate bill. The cost can be significant (several thousand dollars a month for large ads). Once the ad is printed in the directory you cannot cancel for one year!

Be Careful What Numbers You Inadvertently List

It is not uncommon to call Directory Assistance, request the telephone number of a large company, and be given a number that turns out to be answered by a fax machine tone. How does this happen? And what can you do to prevent it?

How it happens is that someone in the organization calls the telephone company to place an order for a new fax line (an outside line with a separate telephone number). If the telephone company business office fails to ask whether this number should be listed, then it will often make it into directory assistance and sometime be given out instead of the main listed number. Or the directory assistance operator may tell a caller, "I have two numbers, which one would you like?" Of course the poor caller (your customer) has no idea. (Tip: If I am the caller and I encounter this, I ask for whichever number has a double zero in it, since if the company is large, it probably has a main number ending in double zero).

To prevent the above scenario, it is important that you establish a company policy so that only one individual (or department) has authority to place orders for telephone service and is responsible for telephone listings. Keep a letter on file with all telephone service providers your company uses and check regularly to be sure this shows up in their records (test it by placing a call to see if they'll take an order from an unknown name for a new telephone line.) Request a written confirmation of all listings from your telephone company on a regular basis (once every 6 months to a year. More if you make frequent changes). You may be surprised at how hard it is to get this , which usually means that no one is paying attention to it.

"Unlisting" Telephone Numbers

To prevent inappropriate telephone numbers from cropping up in directory assistance listings, make sure that these numbers are all "non-listed." Historically, the local telephone company has offered what is called "non-published" service to residence customers. This is referred to as an "unlisted number." There is not an equivalent service for business. Actually every business is required to list at least one telephone number (We're not sure why – maybe some arcane rule from the telephone company tariffs.) Other business numbers not to be listed are designated as "non-listed." The term "non-published" is typically not used for business.

▦ Printed Directories

While it may be hard to find a copy of the printed white and yellow pages in most offices, they still do exist and are actually regularly delivered by hand once a year. The directory business is undergoing a metamorphosis as you will read in the next section, but for the moment at least, these printed directories exist and are used by a fair amount of people. As a business, your concern with these directories is that if your listing information is not accurate in the directory assistance database, it will eventually end up inaccurate in print which cannot be retracted. It is even more critical to check for accuracy in any Yellow Pages Advertising you may be paying for, since this is costly and an error in your telephone number or address can result in much lost business if your customers look here to find you. Each telephone company has a "Closing Date" for the white and yellow pages (usually different dates) after which no further changes can be made.

▦ Keeping Your Listings Accurate As New Telephone Companies And Multiple Databases Proliferate

Before 1984 when the old Bell System was broken up, the local Bell telephone company was responsible for your listings and shared the information with AT&T Long Lines (the long distance division of the Bell System) so databases containing this information were consistent and remained in sync. Today, there are many local and long distance telephone companies. These and other new organizations are getting into the business of preparing and maintaining name, address and telephone number databases. The caller who is calling directory assistance from his cell phone may reach a directory assistance company using different databases than the one he may reach when calling from home. Still another company using other databases may be accessed from the office telephone system.

The point is that there are multiple databases that need to be kept up to date and different companies are accessing different databases (typically more than one) to provide directory assistance service. In addition, many organizations are instructing their employees to use on-line web based directories instead of placing a costly directory assistance call. Directory assistance can cost from 45 cents to 95 cents per call and more for international directory assistance.

The place to start checking your own listings is with the local telephone company who provides the numbers you want to list, whether it is the original local company that has been around for years or one it its competitors (AT&T, MCI Worldcom and other traditional long distance companies now sell local telephone service, along with many others.) Contact the representative in the business office (not your sales person) to review what listings they show for you and determine how they provide those listings to other companies and other databases (you will have to do some digging to get a good answer).

To test how accurate your listings are, get into the habit of dialing 411 and the area code + 555-1212 to ask for your company's telephone number from a variety of pre-selected locations. We suggest calling from telephones in different parts of the country and the world that use the major local and long distance companies. (both the traditional telephone companies for local service and the major competitors, AT&T, MCI Worldcom, etc.). Your local telephone company may have a different method (other than 411) of accessing directory assistance, such as "00" so find out what this is and use it from locations using their service. We also suggest calling from cell phones from different service providers. (AT&T, Sprint, Nextel, Verizon, etc.)

You may also want to check your listings in some of the major on-line databases, often used by the various directory assistance services. These include www.switchboard.com. If you have an 800 (toll free) number that you want listed, you can dial 800-555-1212 to check on this and begin to investigate on-line toll free databases as well.

At some point in the future, we may no longer use directory assistance as we know it today. The unwieldy printed telephone books may become collector's items as it becomes easier for us to access information on-line with portable devices. Nevertheless, in the foreseeable future, people will continue to want the convenience of asking someone for a telephone number (as an added convenience at lease one directory assistance service connects to the company you're calling for no additional charge beyond the directory assistance cost). Having your listing there when someone is looks for you is a basic step for any business.

Your Telecommunications Support Staff

"Help! I need somebody.
Help! Not just anybody!"

– The Beatles, 1968
– Telecom Managers, 2002

Suggestions For Outsourcing Telecommunications Support

People. People who need people.

If you manage your organization's telephone systems and services, your list of responsibilities is getting longer. You may find yourself slipping behind as more bills come in, more contracts come up for renewal and stacks of reports land on your desk for analysis. You install more systems and services, each requiring a different expertise and experience to manage: PBX, Automatic Call Distribution, Voice Mail, Automated Attendant, Converged computer and telephone applications, Interactive Voice Response, Fax Server, Call Accounting and Cost Allocation, Facilities Management, Cabling, Local and Long Distance Services, Internet Access, Wireless Services.

Even the largest organizations are finding it difficult to assemble, train and maintain a staff to manage all aspects of the telecommunications systems and services. The solution may be to look outward to other firms or individuals who can help. The purpose of this tutorial is to provide an overview of the current alternatives for this outside support and to offer some suggestions for making it effective.

Definitions

First, let's create some definitions. Make sure when you're discussing support services, that you and the person you are talking to agree on the terminology.

▶ *Complete Outsourcer*: If you completely outsource your organization's telephone system and services, this means that you have delegated all responsibility to an outside firm including the repair and maintenance of your PBX, provision of local and long distance service, reviewing and payment of telephone bills and a variety of other services that should be carefully spelled out in your contract. The outsourcer is paid a monthly fee and also makes money reselling you equipment and local and long distance service. This type of a relationship may seem like "the fox minding the hen house," since the organization doing the billing is also responsible for approving the bills. Nevertheless, it is working well for a number or large organizations.

▶ *Telecommunications Management Outsourcer*: This type of outsourcing manages your telephone system and service providers, but the company does not sell you equipment, local or long distance service as a Complete Outsourcer may. The Telecom Management Outsourcer is the equivalent of having an in-house telecommunications department, but an outside firm staffs it.

▶ *Out-tasker*: An out-tasker, either a firm or an individual, takes the complete responsibility for one or more specific tasks. For example, you may out-task the updating and production of your telephone directory. Another example would be to out-task the work order processing and record keeping for changes to your telephone system.

▶ *Consultant*: Many people refer to all subcontractors as consultants, so this is one view of the term. In the stricter sense, a consultant is an individual who is charged with analyzing a problem or assessing a situation and providing a recommendation. A consultant may also execute steps necessary to carry out the recommendations. A consultant may be an individual, may have a small company or may be a member of a small or large consulting firm.

▶ *Technician*: Like "consultant", this is a term that has varying definitions. A technician is typically someone who can do the things a purely administrative person cannot. For example a telephone system administrator can write up requests for changes to the telephone system such as adding two new extensions to a particular telephone. The administrator, if trained to do so, can also make program changes to the telephone system to add the two new extensions. If, however, a new circuit board for the PBX was required for those new extensions, or if any rewiring had to be done, that would be the job of the technician.

▶ ***Subcontractor or Independent Contractor***: These term are used interchangeably and typically refer to an individual who is providing work to you for a fee, but is not on your payroll. The fee is usually per hour or per day (also called "per diem"). Depending upon expertise or experience, this person can provide support in any area of your department. Sometimes an entire firm is referred to as a subcontractor.

Any of the above categories may be considered to be a subcontractor or independent contractor.

▦ Resources – Who To Call

▶ ***Telecommunications Outsourcers***: Some firms identify themselves as Outsourcers. At present, they seem to be going after the larger, multi-site businesses. Initially they outsourced the Information Technology systems, but have added Telecommunications to their capabilities.

▶ ***Large Telecommunications Vendors***: Some large telecommunications companies will take on a complete outsourcing of telecommunications equipment and services.

▶ ***Consulting Firms***: Some consulting firms also provide on-going support which may include Telecommunications Management Outsourcing or Out-Tasking in specific areas of expertise. The consulting firm will assess your requirements and may provide management of the Out-Tasking, which may make it a better choice than simply hiring a group of subcontractors from separate firms or from a temporary agency.

▶ **Specialized Support Firms:** Some firms provide support specific to a certain type of product or service or, in some cases, specializing in services relating to equipment from a particular telecommunications vendor.

▶ **Temporary Agencies Specializing in Telecommunications Staffing:** These firms operate like any temporary agency, but maintain resumes on people with different capabilities in the telecommunications arena. Try to find a firm that is truly a temporary staffing firm, rather than a company whose specialty is really permanent placement and only dabbles in the temporary market.

▶ **Temporary Agencies:** Some administrative tasks for managing telecommunications do not require the services of a seasoned telecommunications professional and are trainable. Think about the skills needed for the task and locate a temp agency who can send over some smart, willing to learn, trainable people.

▪ Types Of Management Services Provided

Almost any type of activity for managing telecommunications can be handled using outside support including:

▶ A project such as an office relocation, telephone system acquisition, office expansion, PBX upgrade, addition of Voice Mail, identification of converged computer and telephone applications, etc.

▶ Ongoing record keeping, manual or computer based.

▶ Making program changes to the telephone system or related systems such as voice mail.

- Placement and tracking of telephone system work orders.

- Administration of telecommunications expenses and cost allocation systems.

- Review and verification of monthly telephone bills.

- Preparation and maintenance of corporate telecommunications budgets.

- Preparation and maintenance of corporate directories.

- Liaison with telecommunications vendors on a day to day basis.

- Evaluation of competitive services offered local and long distance carriers.

- Training of users, switchboard attendants or system administrators.

Suggestions For Finding And Managing Outside Support

- The best way to find help is to ask around. Talk to your associates with similar responsibilities at other firms to see whom they have used and what their experience has been.

- Talk to members of User Groups for the specific product or service with which you need help. Most PBXs and related systems have organized User Groups that meet once or twice yearly.

- Ask your vendors or consultants. Most are plugged into the local marketplace and can point you in the right direction if they cannot offer the kind of help you are seeking.

- Define your needs carefully and try to get a person or service closely matching what you need. "Telecommunications" encompasses a vast body of ever changing knowledge. Make sure that the person or company who will provide you the service has up to date and relevant knowledge and experience with the task at hand. For example, many people who know PBXs do not know how to read a telephone bill. Another example is a PBX engineer who understands the technology, but has little knowledge of how organizations use the telephones.

▶ Think about the dynamics of in-house and outside support staff working together. Keep the lines of communications open with both groups. Be careful you don't unwittingly create two different "camps". It is generally best to delineate responsibilities so that the in house staff does not overlap with that of the outside support people. This is not easy to accomplish!

▶ Even with a Complete Outsourcing, it is important to keep some checks and balances in place to monitor the success of the service. Some proposals for outsourcing are very broad. Get specific. What are the tasks and how will you measure whether they are being successfully accomplished? How will the expenses be monitored to be sure the costs are competitive with what you could negotiate on the "open market" rather than buying the services through your outsourcer?

▶ Whether outsourcing, out-tasking or hiring subcontractors for daily help, ask for a work plan in advance, a detailed time sheet as the work is completed and demonstration of results.

▶ Be clear about fees and responsibility for insurance, workmen's comp and payroll taxes. There are many and changing rules regarding the hiring of subcontractors.

The Switchboard Operator

Dinosaur or Company Concierge?

While the casual observer may think that Voice Mail and Automated Attendant systems have replaced the traditional switchboard operator, a growing number of organizations are reintroducing the human touch to handling callers. There is also a changing the view of the switchboard operator's role, with new products being developed to support this.

Before the 1960's, most telephone systems were controlled by cordboards attended by the switchboard operator. Office telephone users relied on this operator to answer all incoming calls, to place an outgoing call, to place an internal call, to transfer a call or set up a conference call and to take and distribute messages. The office telephone did not have the capability to perform these functions as it does today. The operator always knew who was talking to whom on the telephone and how to find people. She was the best source of information of who was in the office, who was out and what time he would be returning. There was also a lot of personal service provided, particularly to the company executives such as screening and announcing callers, placing calls, making restaurant reservations and handling personal business. As telephone systems became more automated the users relied less and less on the operator. The automation also took away the much of the operator's view of what was going on. Calls came directly into telephones, totally bypassing the switchboard, messages were taken by voice mail and the "busy lamp field" which showed who was on the telephone was being phased out by telephone system manufacturers.

The trend toward fewer secretaries combined with the frustration of callers "never being able to talk to a live person" is encouraging organizations to rethink their switchboard operations.

The new role of the switchboard operator includes the following...

▶ Answering callers to the main business telephone number if the caller does not have the direct dial number of an individual.

▶ Answering callers to direct dial numbers that are not answered and forward back to the switchboard.

▶ Transferring callers who wish to leave a message into voice mail.

▶ Answering callers who reach voice mail and press "O" to escape to a live person.

▶ Assisting people within the office who dial "O" for help in locating someone else within the office or in another office, particularly in the larger companies.

▶ Providing up to date information on extensions and telephone numbers within the organization using an company directory.

▶ Providing assistance to company staff on how to use the telephone system functions.

▶ Placing calls for people within the company, particularly executives.

▶ Announcing the delivery of lunches, packages, etc. if the switchboard area also serves as the reception area (Note: While many organizations put the switchboard console at the reception desk, this almost always leads to problems with callers waiting to be answered while the receptionist gives priority to a visitor).

▶ Paging someone over the office paging system while a call is held at the switchboard.

▶ Beeping someone who is out in the field while a caller is held at the switchboard.

While it is clear that the switchboard operator is a customer service department of sorts and a critical component in the organization wide scheme of call coverage, the switchboard consoles provided by telephone system manufacturers do little support this. A typical implementation of a telephone system does not give much thought at all to the set up and programming of the switchboard console (or consoles – most organizations-with over 200 telephones have 2 or more switchboard operators.)

Some of the problems encountered include the following:

▷ As in any incoming Call Center, in a room with 2 or more switchboard consoles, incoming calls should be evenly distributed. This is seldom the case and exactly how one console is selected over another to receive a call is often not clear. Most telephone systems are not flexible in how the system can be programmed to distribute calls to multiple switchboard operators. Some systems claim to send the next call to the console which was idle the longest, but if a caller is put on hold, the system interprets this as being idle. This can result in additional calls coming into a console that already has several calls on hold, while other consoles remain idle

▷ Switchboard consoles are typically not set up to enable calls from different sources to be prioritized (capability to do this at all varies by manufacturer of the system). For example, the mail room clerk calling to see if lunch was delivered by dialing "O" may be answered while an important customer calling the main number waits!

▷ Most telephone systems provide only limited reporting on the telephone call activity at the switchboard. One can often get more information about incoming and outgoing calls from a single telephone extension than from the switchboard console. This lack of information in terms of the volume and type of incoming and outgoing calls and the number of calls on hold makes it difficult to staff for peak periods and to identify which types of calls may be going unanswered or are answered after many rings.

▶ One of the biggest administrative challenges facing today's ever changing organizations is that of keeping the organization-wide telephone directory current. People change offices and many people work in the field and use "hoteling" when they are in the office. People have direct dial number, fax numbers, cell phone numbers and e-mail addresses. Keeping on top of all of these is never easy and many people rely on the switchboard operator to provide the most current information. The challenge is to provide the switchboard with current information.

▶ Even if the directory is kept up to date, there is still the larger problem of the switchboard operator not knowing enough about who does what in the organization to send the callers to the right place on the first try. This results in a caller being transferred several times, perhaps even reaching voice mail and then returning to the switchboard console madder than a wet hen and having a very poor impression of the organization!

▶ We mentioned the "busy lamp field" above as being phased out. Some manufacturers have now phased this useful tool back in. As switchboard operators tend to be isolated from the rest of the office (often relegated to a stuffy windowless room unfortunately) it is useful for the operator to have a visual field of who is on the telephone to provide better information to callers.

▶ Most switchboard consoles have some clunky ways of operating that take more time than is necessary. Supposing an organization has 4 digit extension numbers. If a call comes into the switchboard, the operator must press one button to answer the call, four buttons to dial the requested extension and another button to send the call. That's six presses for every call! Some consoles have what is called a DSS (Direct Station Select) enabling the switchboard operator to press one button to extend the caller to a particular extension and one more button to complete the transaction.

▶ Switchboard consoles have what are called "loop keys" which are the buttons where callers are put on hold. This may also be where new calls come in. There is typically a limited number of these loop keys, so not too many callers can be held by one operator and more calls may tend to get backed up behind these calls on hold until the operator can end the call or send it to an extension.

▶ Another limitation of many switchboard consoles is that the operator cannot see the number of calls backed up. Some consoles show a flashing lamp to indicate a back up, others do actually show the number of calls in the back up, but don't provide any indication of the type of call backed up (i.e. new incoming call, voice mail escapee or internal call by someone dialing "O")

Are there any solutions to improve things? Sure! Do they cost some money? Almost always. So you have to decide what it's worth to your organization to either provide better service internally or to callers. We recommend not looking at the switchboard operation in a vacuum, but rather as a part of the overall plan for handling calls and providing service and support to callers and staff members inside and outside the office.

Some Suggestions For Improvement:

The place to start is to find out what capabilities you have with your existing switchboard consoles. Don't be surprised if you have to be very persistent in getting this information. And validate whatever you are told by getting it in writing or actually testing it on your system. For example, in many systems, with some creativity, you can figure out how the switchboard operator can distinguish between new callers, callers returning from voice mail and internal calls, enabling you to establish priorities. Try putting each of these on separate buttons or having different messages appear on the switchboard console display.

Consider new directory software packages that facilitate keeping the directory current without a lot of manual data entry or separate data entry into multiple directories. This generally takes a company wide effort among different departments including Human Resources, Telecommunications and Information Technologies. Some directories reside on a PC that doubles as the switchboard console and enables the switchboard operator to find a name and send the call to an extension using the "mouse" or the PC keyboard which is usually faster than the mouse. Some of these directory packages also enable the operator to search for the correct person in an organization by a key word that the caller may use when looking for a particular department. This is particularly useful in organizations offering multiple and complex services such as accounting, consulting and law firms.

If you decide to truly view your switchboard area as a Customer Service Department and you have two or more switchboard consoles, consider replacing them with an Automatic Call Distribution system which can be separate from your PBX or a subsystem of the PBX. The advantage of doing this is that it will give you better reporting, tracking types of calls and quality of service. It enables you to prioritize and handle calls more efficiently. Internal callers can hear an announcement that their call is in line to be answered, rather than just hearing ringing. There are some small PC based systems of this type that also enable you to experiment with adding computer capabilities to the telephones. For example, when an executive of the company dials "O" a screen pops up in front of the switchboard listing that executives preference's for restaurants, travel arrangements, how callers are to be screened, etc. Using caller ID a client of the firm may be identified and his information will pop up on the screen indicating who handles the account, recent problems, etc. (Note: If you take this route with an Automatic Call Distributor, find out if your system still needs to have a working conventional switchboard console, some do, and how this will fit into the overall plan.)

You've come a long way, Ernestine!

Managing Telecommunications Expenses

"Can you believe the average corporation doesn't know how much they're spending on telecommunications?"

Planning Your Telecommunications Budget

"How come the cost of a call is dropping and our expenses are rising?"

Why Most Organizations Can't Budget Accurately

Most organizations do not know how much they spend on telecommunications. It is therefore difficult to prepare a telecommunications budget to enable planning for this increasing and elusive expense.

Typically, the first approach to finding out how much is spent is to ask the accounting department to provide totals from the general ledger. This is the part of the accounting system that tracks the detail of all the expenses and income of the organization. Some dollar amounts are usually produced from this request. The problem with using these numbers to plan a budget is that the people in the accounting department do not know how to look at telecommunications bills or how to determine what type of expense they represent. It is rare that the numbers coming from accounting accurately reflect the total spending. It is as common for them to understate the costs as to overstate them, but in either case, it does not support the preparation of a realistic budget.

Even more unlikely is a report from accounting that provides accurate breakdowns as to the type of telecommunications expense (fixed local service costs, variable long distance costs, voice mail maintenance, Internet access, etc.) or information on what purpose in the organization is supported by this expense. (Customer Service center, Sales Force, etc.)

Even when bills are reviewed and approved at the department level, the situation is much the same since the approval takes the form of a signature on the bill saying it is okay to pay it, but not indicating what is being paid for or its purpose.

Organizations with sophisticated systems for charging back the cost of telecommunications expenses to departments are not immune to these issues, since the charged-back amounts often do not bear up to scrutiny in terms of what services and purpose they represent.

Organizations with in-house telecommunications departments may do a little better at estimating true expenses, but it is unusual to find that the expenses for all telecommunications are managed by a single department. Thus, no one individual has the whole picture.

Note: We make the above observations based upon 25 years of experience working with hundreds of organizations on their telecommunications expenses.

Start By Making A List

We suggest starting with no assumptions and building the information from the ground up.

First make a list of all possible telecommunications expenses that your organization may incur, either monthly, occasionally or just once. Here is a list of some major categories to get you started. Individual organizations may have other items to add or may not wish to include all of these categories.

You may want to put the list on an Excel spreadsheet, so that it can become the basis for building the expense record as bills are identified.

Fixed Costs

▶ Monthly costs for outside lines connecting you to your *local telephone service provider*. (part of your local telephone company bill)

▶ Monthly costs for outside lines connecting you to your *long distance telephone service provider* (part of your long distance bill – may be shown as "access charges")

▶ Monthly costs for outside lines *connecting you to another office of your organization.* (may be called "tie lines" although this term is becoming somewhat dated.)

▶ Monthly costs for outside lines for *data transmission to other offices within your organization.* (computer-to-computer communications between two points or among multiple locations)

▶ Monthly costs for *Internet access.*

▶ Costs for the *maintenance contract on your telecommunications equipment* including PBX (telephone system), voice mail and related systems and hardware. (this may be monthly, quarterly or annually)

▶ Monthly cost for cell phone usage (if you have an agreement for a fixed cost covering a given number of minutes)

▶ Monthly rental of pagers and beepers and associated services

▶ Cost for yellow pages advertising and other listings

▶ Salary and benefits for employees responsible for telecommunications management.

▶ Monthly fees for outside services providing telecommunications management.

▶ Cost of space for housing telecommunications equipment

● Variable Costs

▶ Local telephone calls

▶ Directory Assistance

▶ Long distance telephone calls (outgoing and incoming toll free)

▶ Calling Cards

▶ Audio teleconferences

▶ Video conferences

▶ Purchase of telecommunications equipment (telephone instruments, circuit boards, modems, faxes, cell phones, video conferencing equipment etc.)

▶ Installation charges for outside lines (on the local or long distance bill)

▶ Installation charges for telecommunications equipment.

▶ Professional Services/Telecommunications Consulting, Management and Administration

▪ Find Those Bills

Once you have the categories of costs set up, get copies of all bills that are related to telecommunications. Getting a copy of each bill received is not as easy as it sounds.

Often bills are approved and passed along for payment with nothing kept other than the front page of the bill which does not provide information on the bill detail (usually just a total.) Sometimes multi-page bills are pulled apart and never go back together in a coherent manner.

It is typical for different types of telecommunications bills to go to different people within the same organization, so you must approach each person or department to see which bills are received each month. For example, the data communications and Internet access bills may go to the people in the computer room while the local and long distance telephone service bills go to the facilities manager and the cell phone bills go to the office administrator. Pager and beeper bills may go to the security department. Ask each person to save you the complete bill when it comes in (this is easier than finding old bills).

▪ Identify Costs By Category

One complete bill may be lumped into a single expense category or may include different expense categories. For example, the local telephone company bill includes fixed monthly costs for outside lines and variable costs for directory assistance, local calls, installation charges and perhaps long distance calls. Without looking at each of these individually, it is difficult to use judgment for budgeting future costs. You may have added people recently and have installation costs on a bill for additional outside lines. These costs may not necessarily recur next year.

Deciphering most telecommunications bills is difficult, so do not hesitate to call the billing representative (not the sales person) for each company sending you a bill and ask to have the bill detail explained to you in a clear manner, not using technical terms. If anyone uses a term that you do not understand, ask for an explanation. Nothing you are being charged for cannot be explained. Bills can be so complex now that even the companies rendering them can't understand them, so persevere!

Enter the expenses in the spreadsheet under the expense category, referencing who the bill is from and the account number. There are many different ways to set up a spreadsheet. The important thing is to capture the information and be able to arrive at a total.

Sorting Costs By Their Purpose

To really understand the expenses, it helps to know not only what category of telecommunications expense each falls under, but what business purpose is served. For example, of all outside lines from the local telephone company, are some for fax, others for computer modems and others connected to your PBX? Or is one portion of the telephone system maintenance bill for the equipment supporting your Customer Service center as opposed to the Administrative Offices?

Now What?

Once you think you have all bills and have them categorized, total them up. See how this compares to the information provided by your accounting department. Since it is unlikely that the two will match up, begin the process of identifying all bills that make up accounting's totals and see where the discrepancies are. Once you think you've reached a reasonably accurate total expense and the elements that make it up, this can provide the foundation of your budget.

Some organizations just take what they spent last year and add 5 or 10% to arrive at a budget (the planned expense) for the following year. In most organizations, overall telecommunications costs are rising, even though the cost of some services, such as long distance calling, has been dropping dramatically. This is due to increased use including fax and a computer lines and to the addition of cell phones.

As with all budgeting activity, it makes sense to find out what will be happening in each department to identify what may be coming up in terms of growth or added equipment or services. Tracking anticipated vs. actual expenses is important to see if your budget numbers are proving to be accurate.

Other factors may affect the budget numbers. For example, if you're eliminating all of your dial-tone lines used for computer modems and replacing them with a high capacity circuit connecting you to the Internet, your local telephone company bill will go down, but some of the cost will be replaced by the bill from an Internet service provider.

Keeping good records of your telecommunications equipment can also help you to plan your budget more accurately. If you know that your PBX (telephone system) is about to reach maximum capacity, you can get cost estimates for upgrading or replacing it and add that to next year's budget.

You can never have too much information about your organization's telecommunications expenses. Preparing and managing a Telecommunications Budget is a good place to start.

Found Money —
How To Audit Your Telephone Bills

A million here – a million there…
it can add up to real money!

The purpose of a Telephone Bill Audit is to verify the accuracy of the billing. It ensures that you are paying only for services you have in place and are paying the correct amount according to a tariff (a written description of services and costs) or a contract.

Validating Fixed Monthly Charges From The Local Telephone Company

Here is a step-by-step process for auditing the fixed monthly costs from your local telephone company:

1. Get a copy of a recent telephone bill.

2. Look for the first lump sum line item, typically called "Basic Service." It may be on the second page of the bill, behind the page that gives the total amount owed. This amount has nothing to do with the calls you made. It is primarily made up of the costs for the lines (also called circuits) that connect your organization to the outside world. Unless you add or remove outside lines or change your service, this amount will be exactly the same every month.

3. Call the Business Office of your local telephone company (telephone number is on the bill) and ask for a breakdown of the component charges of that lump sum. This may be called a "Customer Service Record" or "Detail for Fixed Recurring Monthly Charges." Ask the representative to mail it to you and find out how long it will take. In some instances, you may be able to access this on a web site if you have the password.

4. When you receive the customer service record, much of what you will see is written in USOC (Universal Service Order Codes) used by the telephone company to identify different types of services. Go through the record and highlight all charges listed and anything that looks like a telephone number or a circuit number. The telephone numbers have 7 digits such as 883-1234 and other circuit numbers may have 2 digits followed by 3 or 4 letters and then 5 or 6 more digits (for example 96 OSNA 34756). Make note of the USOC codes next to each line item with a telephone number or circuit number. It is likely that these have charges associated with them.

5. Next, call back to the local telephone company business office and ask the representative to explain the Customer Service Record to you in terms of what the USOCs mean. For example "1MB" may refer to a single business telephone line and "TTN" is one of the USOCs for touch-tone service. These codes are not uniform and vary from state to state.

   Some circuits may have multiple components including "mileage" for circuits connecting two points.

   Ask for an explanation of each line item and for a validation that the associated charge is accurate. Another way to check the accuracy of the charge is to subscribe to a tariff service. This service provides up-to-date rates for most local telephone companies in the United States.

6. Once you have validated that the rates for each line are correct, verify that all telephone or circuit numbers for which you are being billed are actually in place. You may want to enlist the support of your telephone system maintenance company here. Go into the telephone equipment room and find the "demarc" (demarcation point) where the local telephone company has dropped off your outside lines. Some older demarcs may be labeled with little white tags on which the telephone number or circuit number is written. They can fall off (and do over time!) Newer demarcs may have orange

covers with a place to write the circuit number on the back of the cover. Gently open it up to see if the numbers are written there. What you're doing now is looking for all the telephone numbers and circuit numbers that appear on the Customer Service Record.

Just because you find the number on the demarc does not mean that the line is actually there. It may have been there at one time. That's where the telephone system vendor comes in. Someone needs to use a test set to see if there is dial tone or "battery" on the line meaning that it is a live line. If it is not, you may have the basis for claiming that the line for which you are billed is not active. Or if you cannot find a line that is on the Customer Service Record, you may also have a basis for requesting a refund. Sometimes a line may be there, but is not labeled. Ask the telephone system technician to provide you with a list of all telephone numbers going through your telephone system. You may also have a list of fax and modem telephone numbers. All this information can be compared to the numbers for which you are being billed.

7. Once you have identified incorrect billing or billing for lines you can't find, prepare a letter to your local telephone company business office, outlining your claim. Some telephone companies have special departments called "Accounts Reconciliation" to handle this.

8. If your claim is valid, you will receive a check or a credit on your bill for the amount of the refund. If you have evidence of how far back the incorrect billing went (in the form of old Customer Service Records) or have a letter requesting that a line be disconnected, while the billing continued, you may be able to get credit back to that date. There are different rules in different states, but the average refund gives credit back from two to three years. It also should include all taxes and surcharges that were paid on the incorrect billing. Some states, such as New York, also require that the telephone company pay you interest!

If this is more than you want to tackle yourself, think about hiring a professional auditing firm.

Professional auditing firms know what to look for and may be able to identify things you may miss. It is also likely that they can recover more money than you will, since their fees are based upon a percentage of the refund.

Long Distance Usage Charges

Validating Long Distance Usage charges is another type of telephone bill auditing.

1. Find out what the charges should be. This is not as easy as it sounds. Ask your long distance company representative to help you. If you signed a contract, this may spell out the rates you should be charged, although many do not.

2. Find out the rates for which you contracted for each type of call. There are typically many different rates (cost per minute) depending upon the type of call. Some of the categories include: switched (going out over your regular telephone lines) and dedicated (calls carried on a high capacity circuit connecting directly to your long distance company.); international (different rates to each country), interstate, intrastate and intra-LATA (these are toll calls within your nearby geographic area), outbound and inbound (800 calls), peak and off peak (or sometimes day, evening and weekend), calling card calls (may include a per call surcharge as well as cost per minute) and long distance directory assistance calls.

   Some long distance carriers rates are based upon a percentage off of the tariff rates. Sometimes different percentage discounts are applied to different types of calls (for example a higher discount on outbound than inbound).

   Most long distance contracts do not guarantee these rates for the life of the contract. Thus, there can be rate increases throughout the contract. In order to validate the billing, the percentage of each increase, when it took effect and to which type of calls it was applied, must be known.

3. Once you have all this information together, you can look at each type of call on a representative bill to validate that the cost per minute is as it should be. Another variable is "rounding" so it's important to find out how your carrier rounds the calls (To the nearest minute? Second? Six second increments?)

Some long distance carriers show the actual cost of the call after discount next to the detail of each call. Others show the pre-discounted cost per minute on the call detail and show the discount as a lump sum elsewhere in the bill. Find out how your carrier shows the discounts.

If your bill breaks down total minutes in each category, divide the minutes into the total cost as another way of validating the cost per minute.

Most long distance contracts do not guarantee these rates for the life of the contract. Thus, there can be rate increases throughout the contract. In order to validate the billing, the percentage of each increase, when it took effect and to which type of calls it was applied, must be known.

Traffic Studies –
Measuring Call Volume to Determine
the Number of Outside Lines You Need

"We've designed our system so that every person in our company can be on five different calls at the same time!"

What Is A Traffic Study And Why Do You Need One?

Your organization's telephone system is connected to a number of outside lines on which your local and long distance telephone calls are placed and received.

A traffic study and the art of traffic engineering assists you in determining how many of these lines you need to ensure that callers will always be able to reach you and not get a busy signal. It will also ensure that a line will be available when someone within your organization dials "9" to place an outgoing call.

In addition to supporting good service levels for callers and your staff, a traffic study can prevent you from paying for outside lines you don't need, if you have too many.

The "traffic" refers to the telephone calls. A traffic study "polls" (logs on and requests information from) your telephone system to determine the number of calls and the duration of the calls taking place on each outside line in your system. This is usually done over a period of time such as one week and runs throughout the business day.

If your business is seasonal, be sure to conduct the study during the busiest time of the year. Some businesses actually change the number of outside lines depending upon the time of year, going though the trouble of removing and reinstalling them.

Many businesses are continually changing, so a regular traffic study (every 6 months or yearly) is advisable.

What Traffic Information Is Obtained From The Telephone System?

The raw output of traffic information from your telephone system provides the number of calls handled by each outside line on an hour-by-hour basis. This figure is called a "peg count." (traffic studies of old used devices that knocked down a peg for each call). In addition, it provides the total duration of calls on each outside line, by each hour.

The duration of telephone calls is expressed in "call seconds" using a term called CCS which actually represents 100 call seconds (or 100 seconds worth of calls) The first "C" is the Roman numeral for 100 (so CCS is 100 call seconds). Another term used to represent the duration of calls is an Erlang which is 36 CCS or 3,600 seconds of calls (3,600 seconds in an hour.)

How Do You "Traffic Engineer"?

Once you have the information on the number of CCS or Erlangs in the busiest hour of your traffic study, you're ready to do some traffic engineering (don't try this at home!). There are a number of statistical models which enable you to enter the number of CCS in the busiest hour of the day and the "grade of service" you desire. This grade of service is expressed as, for example, "P.01", meaning that it is acceptable that 1% of all calls will be blocked. "Blocked" means that no outside line will be available for a call to be placed or received. A grade of service of "P.001" means that only one call in 1000 is blocked. The statistical models take this information and derive the number of outside lines you need based upon the call volume and desired grade of service.

This may sound straightforward, but it is not. There are many variables to be considered. For example, the Erlang B model assumes that if the call is blocked the person placing the call does not retry or is routed to another group of outside lines. Erlang B is frequently used to engineer the first group of lines (trunk group) selected when someone dials "9" to place an outgoing call. Another model called the Extended Erlang B assumes that callers who are blocked will retry. It is therefore useful for incoming trunk groups such as direct inward dial and 800 service. The Erlang C model assumes that callers who are blocked will hold on in a queue indefinitely and is more often used for Incoming Call Centers rather than traditional PBXs. There are still other models for situations where calls overflow from one trunk group to another such as with Automatic Route Selection in the PBX, designed to send calls over the group of lines on which calls will be the lowest cost.

Traffic engineering is often done using tables that are based upon the different statistical models. There are some computer programs available that have the modeling capability built in. There are also web sites such as www.erlang.com that provide on- line traffic calculators. In all cases, this work needs to be done by an experienced traffic engineer to ensure accurate and meaningful results.

The end result of the traffic study will be a written explanation of the process, what was determined and the recommendations for making changes, if needed. Some studies also provide graphs to reflect the call volumes distributed throughout the working day, showing the peaks and valleys.

Different Types Of Outside Lines

Another variable has to do with the type of outside lines your telephone system uses. If your system has all "combination trunks," ,also known as "both-way trunks," then the same group of lines handles incoming and outgoing calls and this is straightforward. (Note: A trunk is a single outside line even though it sounds like something bigger. One trunk can handle only one call at a time.)

If your telephone system uses a different group of outside lines for incoming calls (such as DID/Direct Inward Dial trunks) than for outgoing calls (sometimes called DOD/Direct Outward Dial trunks), then the traffic engineering needs to be done for both groups. If you have different "trunk groups" then you will require an overall greater number of outside lines than if the same lines handle both incoming and outgoing calls.

Many telephone systems are connected to either the local or long distance service provider (or to both) with an outside line known as a T1. The T1 handles 24 separate telephone calls. The traffic study can look at all 24 "channels" of the T1, some of which may be designated for incoming calls and others for outgoing calls. If you have a separate T1 to the local service provider and another to the long distance service provider, traffic engineering is needed for each T1. If the T1 is busy and calls overflow to regular trunks, this also needs to be considered in terms of judging the overall number of outside lines.

Still another type of high capacity circuit, like a T1, is the PRI (Primary Rate Interface), increasingly used to connect telephone systems to a local or long distance carrier. The 23 channels of the PRI can be used for either incoming or outgoing calls and do not need to be designated for one or the other. Another advantage of the PRI as it relates to traffic engineering is that the local or long distance service provider can send information to your telephone system over the PRI on how many calls tried to reach you, but got a busy signal, since all of your outside lines were in use.

▦ Who Provides Traffic Studies?

So who do you call to obtain a traffic study? Many local and long distance telephone companies can provide this service. Unfortunately some are dropping this valuable support as staffs are reduced. You can ask you account executive to find out if traffic studies are available, how much notice is required to set one up and whether there is an associated charge. Some local telephone companies like Bell South will place a computer terminal at your site enabling you to monitor the traffic on your outside lines throughout the day. This will only make sense if you are a large organization.

Some telephone service providers use traffic studies as a means of selling more service. If they can show that your callers are reaching busy signals, then you'll need to buy more outside lines.

Find out which model they are using to arrive at their conclusions on how many lines you need to be sure that the number is not overstated.

A major advantage of the traffic studies from the local and long distance service providers is that they can tell you how many callers attempted to reach you and received a busy signal, on an hour-by-hour basis. Your telephone system maintenance company can also provide a traffic study, but will not have information on the number of call attempts that reached a busy signal (unless you have a PRI type of circuit mentioned above and your system is set up to capture this call attempt information.)

While your telephone system maintenance company can almost always provide a traffic study, few regularly recommend it. If they're a small company, they may not have the expertise on staff to interpret the study results. If they do have the needed skills, you may want to negotiate a traffic study into your maintenance agreement. There are some smaller telephone systems that do not provide the output needed for the traffic study.

If your telephone system maintenance company does the traffic study, they can also provide information on other elements of the telephone system that may affect service to callers such as the Central Processor (CPU) Occupancy and the use of "Time Slots" in the PBX matrix. If either of these is overloaded, callers may not be answered promptly on incoming calls and staff calling out may have to wait for a dial tone. These areas would not be addressed by a traffic study from a local or long distance service provider since they do not have access into your telephone system.

The traffic study on the telephone system can also provide information on call volume coming into the switchboard and how quickly callers are being answered. This may be helpful in planning a transition to an Automated Attendant enabling callers to direct their own calls.

Traffic studies can be set up to look at the connections between the telephone system and voice mail to determine whether callers sometimes can't get through to voice mail. They can help to properly size the group of lines used to retrieve voice mail messages for staff members in the field.

If you have what is called a "Call Accounting" system on your premises that keeps track of which extension and which department made each telephone call, you may also be able to set this up to run traffic studies. Check with the Call Accounting system provider. After you obtain the results you may need some support in interpreting them.

What Is The Correlation Between Number Of People And Number Of Outside Lines

Often, people look for a ratio of staff headcount to outside lines. This depends on the size of the organization and the type of activity.

For example, a telemarketing operation may need almost as many outside lines as people. A factory may need very few outside lines. A standard office with 100 people may have 25 outside lines (1 line: 4 people). If you have a small office there will be more outside lines per person (1 line: 2 people). You need fewer outside lines per person if you have 1,000 people (1 line: 8 people). There is no substitute for regular traffic studies to ensure an optimal number of outside lines for your particular environment.

If you're setting up a brand new operation and have no call volume information with which to work, it is typical to overdo it and order more outside lines that you think you need. While there is nothing wrong with this approach, be sure to follow through with a traffic study once you're up and running. Otherwise you may have much more capacity than you need and continue to pay for it, which happens more often than not!

Disconnecting Telecommunications Circuits and Stopping the Billing

"Now where is that letter I sent back in 1993 to have this circuit disconnected?"

Suggestions for Disconnecting Telecommunications Circuits

▷ Be sure of the circuit numbers. Even the service providers sometimes get the numbers wrong, and once a circuit is removed in error, it often takes considerable time to have it reinstated. Part of your request should be to ask the service provider to verify the circuit number before it is removed. They may need to send a technician to either end or both ends of the circuit to do so.

▷ In your request, identify all circuit numbers associated with the circuit. If the service provider from whom you receive your bill buys part of the circuit from another service provider (such as the local telephone company) there may be one or more additional circuit numbers from the local telephone company associated with certain legs of the circuit, in addition to the main circuit number on your bill.

▷ Make telephone contact first with a representative of each service provider. Make sure you are speaking to the right person, which may take several tries. Calling the number on the face of the bill will be a good starting point, unless you have a sales rep who should be able to direct you.

▷ Determine their preferred procedure for arranging a disconnect and follow it to improve the likelihood that your request will proceed smoothly. Once you have determined all of the following by telephone (which probably won't happen without a call back) confirm everything in writing (e-mail will do – but ask the person to confirm their agreement back with an e-mail and print hard copies of the communication for your records. If you decide to prepare a conventional letter of confirmation, fax it and call to be sure it has been received).

- Determine who owns the equipment associated with the circuit including multiplexers, circuit boards, etc. If you own it you may want to reuse it. If the service provider owns it and rents it to you along with the circuit, be sure that it is identified and removed from billing along with the other components of the circuit charges.

- Verify the cost of the circuit so that you can follow up to be sure that the correct amount is removed from monthly billing. Find out where it appears on your bill so that you can follow up to be sure it is removed from billing.

- Determine if there any costs associated with removing the circuit.

- Determine if the circuit is covered by any written contract and if so, what the terms of the contract are relating to the removal of the circuit (suggest doing this before calling to place the disconnect order). If there is any penalty associated with removing a circuit before the contract is up, this penalty amount can be negotiated if you have other circuits with the service provider.

- Do not send the service provider information on the circuits being disconnected from the other service providers (as may appear on your in-house records), as this may confuse things.

- If any part of the circuit to be removed is associated with any new circuit (as is sometimes the case when a circuit is relocated), be sure that the new circuit will not be affected by the disconnect request.

Summary Of Information To Be Included With The Written Disconnect Request:

▷ Name, title, company name, address, telephone number and e-mail address of the person with whom you have spoken to request the disconnect.

▷ Your name, title, company name, address, telephone number and e-mail address.

▷ Confirmation of the date you requested the disconnect by telephone.

▷ Account number associated with the bill upon which the circuit appears.

▷ Main circuit number and associated circuit numbers.

▷ Type of circuit.

▷ End points and any intermediary points of the circuit including street addresses, demarc locations, port designations and all associated terminating equipment information that you show on your spreadsheets (confirming first, with the service provider, that it is all correct).

▷ Date for disconnect and date on which billing is to be stopped. Often you can request a "bill to" date that will stop billing immediately even though the actual disconnect date may be further away (as long as you no longer use the circuit).

▷ Order number (this can be provided by the representative – probably on a call back after your initial conversation, but we suggest waiting for this number to put it in the confirmation).

▷ Arrangements for any premise visit to remove equipment or other elements of the circuit. Note: If the service provider does not make a premise visit to remove the circuit, we recommend removing the label from the location where the circuit was on the demarc – once the circuit has been removed from billing – to keep your demarc up to date. You may also wish to reuse the demarc location and can keep track of this location as being "spare."

▶ Access information if a premise visit is required including name and contact telephone number of person who will arrange for the service provider to get into the space to remove the circuit or associated equipment (date and time if this has been arranged as well).

▶ Identification of all associated equipment, port location, demarc location, etc. if know (all items on your in-house spreadsheet, as long as they have been verified by the service provider first).

▶ Cost of the circuit.

▶ Any arrangements that have been agreed to by telephone.

After the circuit was to have been removed from billing, check your bill to be sure the charges have stopped. This may take several months, but a credit back to the disconnect or "bill to" date should be reflected when the billing finally stops. Check another bill after several more months have passed as charges sometimes turn up again. Look for stray charges that do not appear to be associated with any circuit. Sometimes certain components of the circuit billing are removed and others remain.

Understanding the Nuts and Bolts

"Okay, it's a little more complicated than nuts and bolts — but nothing that can't be explained."

Your Telecom Equipment Room – The Grand Tour
Part One: The "Demarc"

"And don't be afraid of the dark."

Go ahead. Don't be afraid. Crack open that door and step inside. Is it dark in there? Flip on the light switch. Close the door behind you. Leave the office world behind as you venture into "The Telephone Room." It's probably nice and cool and you hear the comfortable sound of humming from power supplies. Red and green lights are flickering or flashing on the displays of different pieces of equipment, indicating telephone calls in varying states of progress coming into or heading out of your offices.

Locating The Demarc

Look around. You probably see a wall covered with a piece of 3/4 inch plywood. Mounted on that plywood you see a group of white or gray plastic objects that have cables running into them. These are called "blocks" (more specifically: M66 blocks or 110 blocks) and may also be called RJ21Xs. Look more closely and you'll see that the larger gray cables have been split open to reveal a group of small copper wires, each coated with plastic of 2 different colors (like orange and white or blue and white). Each wire is associated with one other wire of a complimentary color scheme. The two together are called a "cable pair." The colored wires are connected or "punched down" onto the block. Look back at the larger cable they're coming from and try to follow where it goes. You probably see it joining with other cables that eventually lead to an open metal pipe about 4 inches in diameter (or group of pipes) in the ceiling. You may see similar cables heading in another direction over towards one or more large metal cabinets (beige or gray). Those cabinets are the heart of your office telephone system (the PBX).

This group of plastic (also partly metal) blocks is collectively known as the "demarc" (short for demarcation point.) This is the point where the telephone lines that connect your organization to the outside world are delivered by your telephone service providers (local or long distance company) and connected to your office telephone system which is, in turn, connected to every telephone on every desk. Some of the outside lines may not go through the telephone system, but may be used for fax machines or computer modems.

A regular single outside line uses a single pair of copper wires. A high capacity circuit called a T1 (delivering the equivalent of 24 outside lines) may use two pairs and also requires a separate electronic device to be mounted nearby.

The Telco Demarc

You see an area of the demarc that seems separate from the rest and may see little white paper tags hanging down from it and orange colored snap-off covers over each of the blocks. This is the "telco demarc" (short for telephone company) and is the place where the local and long distance telephone companies drop off their outside lines (also called circuits). The snap-off cover can be opened from one side and swings back. On the inside part of the cover, you see a series of telephone numbers written with a felt tip pen or ink. Each of these numbers is next to the exact location (pair of wires) on the block where the circuit is located. The paper tags are an older method of keeping track of circuit locations, but may fall off over time so the "orange cover" method is better, but still not great. If a circuit number changes or you reuse the cable pair for a different circuit, it is hard to change the circuit number written on the inside of the orange cover, so the information here often gets out of date or is hard to decipher. Or when a circuit is disconnected, the circuit number is often not removed from the inside cover.

Demarc Record Keeping

It is a good idea to keep a separate record of where every circuit comes into your premises. If you don't and the circuit needs to be repaired, it will be time consuming to find it.

A circuit number may be a 7-digit telephone number like 883-1234 or may be a combination of numbers and letters such as 96PLNA43261.

Another reason for keeping records is to manage the cable so that you know where you have "spare pairs" on which the telephone company can deliver additional outside lines. Otherwise you may pay to have a new cable installed, not realizing that you have spare pairs in the existing cable. (A cable most typically has either 25 or 50 pairs.)

Still another reason is to keep a record of the location of any spare outside lines. If you don't, you may order new lines and keep paying for the one's you've lost track of as well.

You can keep manual records in a binder, but we suggest setting the demarc records up on a computer for easier updating. Assign a number to each block. Often the telephone company will have already written on the outside of each cover, sometimes "RJ#1, RJ#2, etc. Each block has room for up to 25 outside lines (unless a particular circuit takes up 2 pairs of wires). So you can identify each pair by indicating the block number and then numbering the pairs from 1-25. (Block #2, Pair #23 or just 02-23). Next to each pair write the circuit number associated with that pair or indicate "spare" if it is not in use. You may also want columns for "date installed," "use of the line" (fax, PBX, modem, etc.) and "service provider". An Excel spreadsheet lends itself nicely to demarc management. Make sure you date the spreadsheet and post it on the wall next to the demarc as well as keeping it in a separate binder. Don't throw away old spreadsheets as they may provide useful history when looking for a circuit on trying to reconcile circuits with the bills you receive for them.

Getting The Information

Unless you are moving into a new space and can start from scratch, you need to request that technicians from your telephone service providers pay you a visit to help you to set these records up by identifying and validating what is currently on their demarc. Expect to pay for the time and ask what the hourly rate will be in advance. Then when you order a new line, advise the person taking the order on the block and pair number where you want the line to appear (ask your service provider for the exact terminology to use to be sure you get what you want as this may vary geographically and by company.)

Try to group similar types of lines used for a similar purpose in the same area of the demarc and leave spare pairs for growth. (All fax lines for example or all lines going into the PBX).

Your high capacity T1 type circuits may not appear on the demarc, but may be run to a separate device (sometimes called a "smart jack") near the demarc. It is important to identify these circuits as well and track their exact location on the plywood backboard, along with the other record keeping information requested.

Another point at which these records become critical is if you decide to replace your PBX. We've seen companies totally lose telephone service and have a hard time getting it back when a technician removed a smart jack and lost track of the location cable pairs that were feeding into it. Don't assume that your service providers or PBX maintenance company is keeping track of all of this – they're not!

Getting back to those gray cables heading up into the pipe in the ceiling… the ones coming from the telco demarc go to some central point in your building and then out underground or overhead, eventually connecting your premises back to their own switching equipment that connects you to the rest of the world.

The Internal Demarc

The larger, but less complicated, area of the demarc is the internal demarc sometimes called the Main Distribution Frame or MDF. If your company has over 1,000 telephones, this may be on a free standing rack in the middle of the telephone room, with "cable trays" overhead, rather than mounted on the wall. Cables coming into it connect to the separate telephone instrument on each desktop in your office. Each of the 25 locations on the block is for one telephone (most telephones use a single pair of wires). As with the telco demarc, each block should have a separate number (#1, #2, #3, etc.) and each of the 25 locations on the block a separate number. The same identifying number should be on the jack into which the telephone is plugged at the desktop. Do not use the actual telephone extension numbers to label the demarc and jacks at the desk, since these will change over time.

The internal demarc does not have snap-on covers and is typically labeled directly on the block where a small space is available for writing. A spreadsheet also can be used to keep track of this demarc. For help in figuring out the internal demarc, spend a morning with the technician from your PBX maintenance company and again, be prepared to pay for this time. On this spreadsheet you can also track the extension number and location of each telephone, along with the circuit board and port number in the PBX associated with that telephone. You may want to consider purchasing Telemanagement Software that is designed to help you to manage this, particularly if your company has more than 1,000 telephones.

The cables going into the pipe in the ceiling from the internal demarc eventually lead to smaller cables connecting to each individual telephone. Other cables coming out of the internal demarc lead to the PBX. Within the PBX the individual telephones are connected to the outside telephone lines.

See Part Two of this Tutorial for the second leg of the Grand Tour.

Your Telecom Equipment Room – The Grand Tour PartTwo

Leave the office world behind you.

This stroll through the telecom equipment room will introduce you to what is really going on in there and why you need to know. In Part 1, we talked about the demarc, which is the end point of the voice communications cabling to the telephones at each desk and to the outside lines connecting your organization to the rest of the planet.

The PBX

PBX stands for "Private Branch eXchange," which refers to the telephone system switching equipment that connects your telephones to the outside world.

The PBX is probably the largest metal cabinet in the telecom equipment room (except for maybe the air conditioner - try not to confuse the two in front of your boss). If your organization is small (under 50 people), the PBX may be mounted on the wall; otherwise, it sits on the floor. The PBX may be one or more refrigerator-sized cabinets or may be a set of smaller stacked metal cabinets. (The very newest PBXs - not yet widely deployed - may look the same as your computer network servers.) Look for a name such as Lucent (Avaya), Siemens, Northern Telecom, Mitel, NEC, or another PBX manufacturers' name on the outside of the cabinet.

The cables running from each telephone in your organization ultimately end up connected to the PBX, as do the cables running from each outside telephone line. The PBX is also known as "the switch," and the switching matrix (hardware controlled by software) connects outside callers to the telephones in your office, as well as gives an outside dialtone to someone in the office dialing "9."

You may want to get some help from your telecommunications maintenance technician to remove the front panels from the PBX cabinets to look inside.

There, you will see one or more shelves with slots along the bottom, into which printed circuit boards have been inserted. (You'll probably also see empty slots, which are spares.) These circuit boards (or "cards") are needed for both the telephones and the outside lines to operate. There are different types of circuit boards, including:

"Station" Circuit Boards

These control the telephones (also called "stations") on each desktop. The circuit board has a specific number of "ports" or electronics places, each associated with a specific telephone. Typically each of these circuit boards controls eight, 16, or 32 telephones. It is important to know which port is associated with which telephone and which ports are unused. Since a circuit board may cost up to $3,000, keeping track of your spare ports will save money, preventing you from buying new boards unnecessarily. Your telephone system maintenance company will not, as a rule, keep these records for you. They can, however, obtain the information on which telephone is associated with which port on which circuit board, and identify spares, using the System Administration Terminal. This is usually a PC which sits on a desk somewhere in the room.

There is a numbering scheme to identify each port such as 01-04-08, representing the eighth port on the fourth circuit board on the first shelf. Find out which system is in use in your PBX. In Part 1 of this tutorial, we recommend using a spreadsheet to keep track of which telephone at which location is associated with which cable number. This same spreadsheet can be expanded to correlate the telephone to its associated port in the PBX.

The two different types of station circuit boards are referred to as digital and analog. In most systems, most of the telephones from the manufacturer of the PBX are digital. If you have some of the older, single-line telephones, or have extensions from the telephone system in use for fax machines, these use the analog type of circuit board.

"Trunk" Circuit Boards

A trunk sounds like something big, but it really refers to just a single outside line that can handle a single telephone conversation.

There are "both way" (also known as "combination") trunks that handle incoming and outgoing calls. There are also DID (direct inward dial) trunks that handle incoming calls only, enabling callers to dial directly to a specific person's telephone within your organization. Some PBXs mix DID and both-way trunks on the same circuit board and others require a separate circuit board for each. As with station circuit boards, each trunk circuit board may handle eight, 16, or 32 separate trunks. Part of record keeping is to identify the actual telephone number (also called the circuit number) of a trunk and associate it to the port, slot, and shelf in the PBX.

"T-1" or "PRI" Circuit Boards

While most PBXs still have some trunks for backup, it is increasingly common to have the outside lines for incoming and outgoing calls brought in on a high-capacity circuit called a T-1 or PRI (primary rate interface: a T-1 with advanced capabilities such as delivering the calling number). These circuits enable 24 (or 23 in the case of the PRI) conversations to take place over a circuit that comes into your demarc, taking up little more space than a single trunk (actually, it requires four little copper wires instead of two). The T-1 circuit board in the PBX sorts the stream of signals coming in over the T-1 back into 24 distinct conversations. A T-1 circuit board may handle one or several T-1s (or PRIs).

Other Circuit Boards

There may be other types of circuit boards in your PBX cabinet too, such as DTMF (dual tone multifrequency, also known as "touchtone") boards, necessary for dialing out and receiving the touchtone signals dialed by each telephone.

There are also other types of circuit boards needed to control the PBX functions.

The point is, that it is important to keep track of all ports, circuit boards, slots, and shelves in your PBX. It helps with troubleshooting, planning for growth, and controlling your costs.

■ Voicemail

In some PBXs, the voicemail system is housed on one or several circuit boards within the PBX cabinet, but it is more common for the voicemail system to be a separate, smaller metal cabinet with its own separate PC-based administration terminal, independent from the one used to administer the PBX.

The voicemail system circuit boards control the system functions and also include circuit boards with ports. The number of ports in a voicemail system represents the number of callers who are simultaneously using the system, including both those leaving and retrieving messages. A voicemail system often includes an automated attendant capability which enables callers to direct their calls. It sounds something like, "If you know the extension of the person you are calling, you may dial it now. For Sales, press 1; for Service, press 2…" While a caller is using the automated attendant, they're also exclusively using one port. Voicemail systems typically can be expanded in increments of two or four ports. As with the PBX, it's important to know how many ports you have, and how much growth is possible; unlike the PBX, voicemail ports are dynamic, and not associated with any particular person's telephone extension or a specific outside line.

■ Call Accounting

Still another occupant of your telecom equipment room may be the call accounting system. This can be a useful tool in managing your system and your business telecom expenses, but is often neglected and out-of-date. Your system is probably PC-based and sitting near the system administration terminals for the PBX and voicemail systems. The call accounting system collects information from the PBX on the calls placed and received, and creates reports by extension and by department. It can track expenses incurred for calls by each extension and department, and enables organizations to monitor abuse (personal calls, etc.) and charge costs back to departments (or clients, for example in a law firm).

Another seldom-used capability of a call accounting system is the monitoring of traffic (incoming and outgoing calls) on each outside line connected your PBX. Looking at the call volume of the busiest times enables you to determine whether you have too many outside lines or perhaps whether it is time to order more. You may see that all outside lines are busy at times, indicating that callers are reaching a busy signal when they dial your number.

Batteries

Some telephone systems are equipped with batteries that keep the system working in the event of a power failure. Backup power batteries, likely on the floor somewhere near your PBX, are also known as a UPS (uninterruptible power supply). It sits between your phone system and the wall socket and may or may not offer surge protection. A UPS with surge protection can save you the cost of lost business and/or a new telephone system.

Boxes Of Spare Telephones

Most telecom rooms include a big cardboard box into which spare telephones and parts of telephones have been tossed. This is not a good system. If a telephone works, but is not needed, it should be tagged or labeled in some way as a working telephone, noted on an inventory, and stored in a locked closet. Telephones that work with a PBX can cost up to $400 apiece, so leaving them lying around or buying new ones when you have spares doesn't make sense.

If telephones do not operate, they should be sent back to the manufacturer for repair or replacement. Check your maintenance agreement to see how this has been arranged with your telephone system maintenance company.

Records

Most telecom rooms have records (old binders, lists attached to the wall, etc.) that are hopelessly out of date. If you take the time to compile accurate records, make sure there is a procedure in place to keep them current. Don't make the telecom equipment room the only place that the records are kept. Keep a spare copy somewhere that another person on staff knows about. (Your phone system may need service while you're not there.)

That concludes The Grand Tour. We hope you have enjoyed the adventure. Now, as you re-enter the "office world," don't forget to turn off the light and lock the door.

Your Voice Mail System

*Yes, it's like a giant answering machine —
but there's more to it than that!*

"Thank you for calling Newcastle Communications. If you know the extension you want, you may dial it now. For sales press 1, for service press 2 or for a directory of extensions press 4." When you hear this you are listening to an Automated Attendant.

If you go into the directory and find out someone's extension and select it or if you dial into the person's telephone directly you may hear "This is Gioia. I'm not at my desk right now. Please leave a message and I will call you back." That's Voice Mail.

Automated Attendant and Voice Mail (AA/VM) capabilities most typically reside within in the same system. The system may be housed on circuit boards inside a PC which is dedicated to the AA/VM or on circuit boards within a telephone system (PBX).

The size of the AA/VM system has two primary variables: (1) the number or ports and (2) the amount of storage time for announcements and caller messages. The number of ports determines the number of people that can be using the system at the same time. These people may be callers listening to the Automated Attendant, callers leaving messages or system users retrieving messages. If the ports are "dynamic" the same port may be used for any of these functions. Some older systems require that you designate specific ports for specific functions. Even if your system has dynamic ports, you may want to think about dedicating some for a specific use. For example, more ports for callers and fewer for system users retrieving messages. AA/VM systems are sold with 2, 4 or 8 port increments. The cost of adding ports to a system can vary dramatically, with 4 port cards costing from $2,000. to $10,000. or more depending on the system. Find out up front how many ports your system can have and what the cost of adding them will be to avoid surprises. If your system operates successfully, more and more people will want to use it, so it is likely that expansion will follow.

Ports may also be required to connect your AA/VM system to the telephone system. There will be corresponding ports in the telephone system to connect it back to the AA/VM system. There are lots of variations on this so ask your suppliers what the options are and what works best. Most AA/VM systems can connect to the telephone system using analog ports (the same type of ports used if you wanted to add basic single line telephones to the telephone system.) Many can connect using some type of digital connection. The AA/VM may connect to a PBX using the same type of port as the PBX would use for a proprietary digital telephone. Some AA/VM systems can connect to the PBX using a T1 (or PRI) connection that provides higher capacity and more advanced capabilities for the 2 systems to communicate. Conceptually, it is important to remember that when a call is in the AA/VM system, it is still using up a port in the telephone system and an outside line. Another consideration is if callers go through the AA/VM system to computer-based applications connected to the AA/VM, is the port released for other uses or does it remain tied up?

The capacity and cost for system storage of announcements and caller messages is a function of the size of the hard drive. As the cost of PCs has dropped, the cost for storage has followed. Most systems now come with 100 hours of storage whether you need it or not. Again, the costs can vary considerably from supplier to supplier with no apparent difference in what you are buying, so it is important to shop around and ask a lot of questions.

Many systems have some relationship between ports and storage that need to be explored before you purchase the system. Can all ports connect to all stored messages? For example, if you have different applications to call into, using different telephone numbers, and the demand peaks for one particular application, will that application be able to take advantage of all ports and all storage in the system?

These Are The Types Of Things You Need To Find Out Up Front

If you already have an AA/VM system operating, and are replacing it, try to get statistics on the number of callers at the busiest times, on caller demand for different resources of the system (e.g. AA vs. VM) and on the length and number of announcements and caller messages you wish to store. If your system is new, make an educated guess or ask your vendor for some guidance based upon others using the system in a similar manner.

Even more importantly, put together an "in English" clear description of how you expect the system to operate. When putting together a written statement of your AA/VM requirements (you've got to do it!), include a flow chart that shows what caller experiences and options will be under each circumstance. Also script out what the announcements will say at each juncture in the system to be sure that there will be sufficient announcement time and ports to accommodate your expectations. The flow chart is particularly important when you are linking to computer applications. The more clear direction you give your suppliers, the better the system will operate. Try to be creative so that your system does not sound like everyone else's. Don't let your staff say "I'm either on the phone or away from my desk." on their voice mail greeting. It's getting old!

Think about using a professional announcement service to record your AA greetings. Be sure the system you are buying lets you customize all announcements to provide you with flexibility in terms of how you use the system.

Since your AA/VM will be linked to a telephone system and perhaps some computer applications (running on separate PCs) it is important to understand how it will all integrate together from both a hardware (e.g. number or ports and type of connection), software (e.g. what do you tell the system to do with the caller when all ports are busy), administrative (e.g. can a single person learn to administer all systems or do you need to hire a staff with special expertise) and practical (e.g. what will callers hear and experience.)

Some converged computer and telephone applications that may link to an AA/VM system include the following:

▶ **Screen Pops** – The caller may be leaving a Voice Mail message and then decide to exit to a live person. If a screen pops up in front of that person who answers the call, what information will the person see? Will it be apparent that the caller was in the VM system? Will the "Caller ID or ANI" (telephone number of the calling person) be carried through the voice mail and enable the computer network to display it on the screen of the person answering? Or is the telephone system (PBX) responsible for holding the calling number and tracking the call.

If the caller has entered some type of identification number when they reached the AA, will that ID number locate a record from your customer database and display it on the screen when the caller leaves the AA/VM system to reach a live person?

What if the caller came through the AA system, elected to use your IVR (Interactive Voice Response – obtaining information such as your bank balance using a touch tone telephone) and then came back to the AA/VM to direct his call to a live person. Will the system maintain the link with the computer network to carry the information on what the caller was trying to do in the IVR, following the caller back through the AA/VM system and the PBX and then display it on the persons screen. "Mr. Smith, I see you were using our automated system to see if a check for $42.11 has cleared your account, let me look that up for you." Without a lot of planning and system integration this won't happen.

▶ *Unified Messaging* – Many AA/VM vendors are now selling what is called Unified Messaging. This enables system users to see voice mail messages on the computer screen along with e-mail messages and faxes. The idea is to have all forms of messaging concentrated in a single place. You can listen to your voice mail through the PC (even the laptop if you're traveling) and respond back. If you're planning to experiment with unified messaging, we suggest rolling it out slowly to a small group of people within your organization so that you can see how it applies to your way of working. If you use "live" call coverage, having people answer the telephone during the day, but transfer callers into the voice mail to leave messages, will the Unified Messaging receive the telephone number of the caller or will it be lost when the secretary answers the call? Find out, it may be lost. Will the caller message show up in the secretary's mailbox as well as the bosses? It may. It's happened.

The biggest question you can ask when implementing an AA/VM system to work in a converged environment is in what similar projects the vendor has participated. Almost all convergence projects require the coordination of multiple vendors and many different skill sets. Identify a project manager and find out what skills are needed both for implementation and ongoing management. Decide who will provide them and factor this into your budget.

Realize that no two AA/VM systems work in the same way. Get to know the ones you are considering. Ask to listen to demos and real systems set up in the way that you plan to use it.

Find out which parts of setting up your applications will require customized software programming and what the variables in cost will be in this area.

Ask the vendors to collaborate on a disaster recovery and ongoing monitoring plan and to put it in writing.

Ask about the strengths and weaknesses of each system. Some systems were originally designed to be an Automated Attendant and are very strong for automated attendant but not for voice mail. Some products also have some queuing capabilities that enable them to work with along telephone systems with Incoming Call Centers.

Most newer PC based telephone systems (Communications Servers) have the Automated Attendant/Voice Mail functions built in as part of the system. But just because these are the latest systems, don't take capabilities for granted.

Users like to purchase systems using common platforms with which they are already familiar. This clearly makes for easier administration, but not all systems use the same platform, so the ease of administration tradeoff may be needed to get the functionality you like from a particular system.

When purchasing an Automated Attendant/Voice Mail system to work with computer-based applications, don't lose sight of the more traditional functions we have come to expect in the past 15 years. Some manufacturers improve systems technically but drop needed functions. For example, when you listen to your voice mail messages will you automatically hear the time and date the message was left. If you don't, this can be confusing, but one manufacturer has eliminated this convenience. Can you speed up or skip messages when you're retrieving them. Can callers replay their own messages before sending them? Do you want them to be able to, tying up your system resources? Make a list of all system functions and decide which are important to you.

Evaluating Your Buildings Telecommunications Capabilities and Service Providers

Find out what your landlord doesn't know and is afraid to ask!

Background

It wasn't too many years ago that you could move your organization into new office space and be reasonably comfortable that the local telephone company would deliver all of the outside lines for telephone service you needed in a relatively short time. "Facilities issues" such as whether there was enough cable running into your building or up to your floor to deliver all the service you wanted was their worry – not yours.

Now that there is competition in the local service marketplace, things have changed. The original local telephone company is no longer guaranteed that they will get the business of every tenant in an office building. Therefore, they are not spending thousands of dollars running new cable either into a building or into a premise in anticipation of the requirements of a new tenant as they would have in the past.

Meanwhile, the new competitors to the local telephone company are clamoring to get into office buildings, hoping that a presence there will be the first step in generating revenue from the tenants or will make them more attractive when they are acquired by another company. Some of these companies are offering revenue sharing deals to the landlords as an enticement to let them in. Building owners are now being faced with a confusing array of options for equipping their buildings with tenant telecommunications facilities.

So what does this mean to you? Whether you're moving to new office space, expanding or just planning to add telecommunications services that you need to remain competitive – better check out your building and see what can be delivered, who can deliver it and how quickly.

■ Checklist Of Telecommunications Services You May Need

The following is a checklist of services you can submit to your building owner. The point is to determine which of these services you can get, which alternative service providers can deliver them and who are the service providers who already have a presence in the building.

- **Local Telephone Service**
  - ▶ POTS lines (plain old telephone service)
  - ▶ Both way lines for PBXs
  - ▶ Direct Dial Incoming Lines to PBXs
  - ▶ Dedicated Circuits to a Local Telephone Service Provider

- **Long Distance Telephone Service**
  - ▶ Incoming toll free calls
  - ▶ Outgoing domestic calls
  - ▶ Outgoing international calls
  - ▶ Calling Cards
  - ▶ Dedicated Circuits to a Long Distance Telephone Service Provider

- **High Speed (High Bandwidth) Data Communications Services**
  - ▶ Frame Relay Service

- ISDN for Video Teleconferencing or other use
- Dedicated high capacity circuits: T1, T3, OC3 etc.

High Speed (High Bandwidth) Internet Access

- Shared high capacity T1 giving smaller tenants access within a few days of moving in.
- Dedicated T1 circuit for exclusive tenant use.
- DSL (Digital subscriber line) circuit.

Access to Internet based services:

- E-Mail
- E-Commerce Retail
- Business to business E-Commerce
- Web Site Hosting

What To Look For In The Building's Telecommunications Infrastructure

- High capacity, high speed fiber optic and copper connections coming into the building.
- High Speed Copper and fiber optic riser cable.
- High speed copper horizontal cable into the tenant spaces.
- Clean, well-organized equipment rooms for housing electronics to support these services. Clearly labeled cables and cable distribution frames. Room for growth.
- Up-to-date, well maintained electronic equipment (different types needed to deliver different telecommunications services)
- Access for running more cable as needed or as technology changes. (Very Important)

> A building owner whose representative can clearly explain the building telecommunications capabilities to the tenants.

Note: Some building owners own and control the cable that feeds into the tenant spaces. Find out how this works and if the building owner will sell access via that cable to any telecommunications service provider you choose.

● What Else To Look For

> Choice of Telecommunications Service Providers

> Fast turn around time for service delivery. Some buildings have this with "on site" provider whose service is already waiting in the building.

> Flexibility to order services from any service provider with whom you choose to do business. Some building owners won't let some service providers into the building.

> Reliability -Telecommunications Services provided by companies with sound financials and a good track record.

> Information - With telecommunications technology rapidly changing, you will need help in determining your current and future requirements.

■ Types Of Companies Who May Deliver Telecommunications Services To An Office Building

There are so many different permutations in terms of what telecommunications companies offer and how they deliver the service, it is a challenge for the buyer to sort out who sells what. It is important to remember that conditions vary from one geographic area to another. Large cities tend to offer more options than remote locations.

The Original Local Telephone Company (ILEC)

In many cases, this company was part of the Bell System before it was "broken up" in 1984. While these companies are merging and being acquired, they are still identifiable in and still provide the lion's share of local telephone service in most areas. Verizon, Bell South and Southwestern Bell (now called SBC) all have a division that is the ILEC (Incumbent Local Telephone Company) who used to be the only game in town. These ILECS tend to come under more government regulation (usually the state public service commission) than their competitors and for that reason tend to be slower to deliver and less flexible in the way that they provide service. Since they used to have a monopoly on local telephone service, this also affects the mind-set of many employees and therefore the delivery of the service. In their defense, no other company has more experience in delivering and supporting local telephone service which is not an easy business as the new competitors are finding.

The ILEC owns most of the cable that runs into most buildings and now has a division that is required by law to rent these cables to the competitors, the CLECS. So ironically, many competitors to the ILEC can't exist without it!

The ILEC sells traditional local telephone service, Internet access (often with DSL circuits) and is now moving into the long distance arena as well so distinction between the two (local and long distance companies) is beginning to go away.

In many cased, the ILEC also owns and controls at least some of the riser cable within a building from which service into tenant spaces is fed.

A Competitive Local Telephone Company (CLEC)

As mentioned above, it is likely that the CLEC rents cable into a building from the ILEC. There are exceptions to this in some instances where there is another company with its own cable (usually fiber) run into certain buildings that then rents this to the CLECs for delivery of service.

Like the ILEC, CLECs sell a variety of services, but tend to specialize in local telephone service. They sell long distance service and Internet access as well. While companies such as these lack the long track record of the ILEC, they tend to be more creative in the services they deliver and sell similar services at considerably lower prices.

A "Wireless" Competitive Local Telephone Company

Winstar was an example of a company who sold local telephone service (it was a CLEC), sometimes using wireless microwave technology to do so. They did not have to rent the cable from the ILEC to deliver their services. They still needed a cable to run from the microwave dish located on the exterior of a building into the tenant space.

A "Long Distance" Telephone Company

The two largest long distance companies AT&T and MCI Worldcom both sell local telephone service in most areas. AT&T bought this capability by acquiring TCG (formerly known as Teleport Communications Group) and MCI Worldcom did the same by buying MFS (formerly known as Metropolitan Fiber Systems). TCG and MFS were early competitors in the local telephone service marketplace. In many cities they did run their own cable into some office buildings, since this was before the ILEC was required to become a "wholesaler" and rent its own cable to its competitors.

An "On-Site" Telephone Service Provider

The concept behind this business is that a the "on site" company makes an agreement with the building owner to bring telephone service into the building, install needed electronic equipment and cable the building with either copper or fiber optic cable or both. These companies are typically found in buildings with multiple tenants with office spaces ranging up to 25,000 square feet. Beyond this point, they cannot compete with the rates for telephone service negotiated by very large companies, but for the small and medium size company they provide the advantage of a single point of contact from which to buy local, long distance and Internet access services. Since they already have some services in the building, they can often provide telephone service with a much shorter turn around time. They typically share a small portion of their revenue with the building owner and may request an "exclusive" in exchange, meaning that no other on site company can come into that building. Check this company out with other tenants using the service if you decide to give it a try.

A Cable Television Company

The cable TV companies have moved into the telecommunications arena. Cable TV was not typically in place in an office building, however, so most of the activity here is still in the residential marketplace, just starting to move into office buildings.

The advantage the cable companies have is that their coaxial cable has very high capacity. The disadvantage is that the electronic equipment connected to that cable needs to be changed to enable the delivery of telephone services and Internet access.

An "Internet Access Only" Company

Some "on site" companies specialize in Internet access only, which is actually the most profitable part of an on-site service provider's business. So if you are a small to medium sized company and want to buy your local telephone service from one company, you may still want to look at this alternative for Internet access.

● A Shared Tenant Service Company

The "Shared Tenant" concept did not take off in a big way, but still exists in some buildings. The Shared Tenant company not only wants to sell you local and long distance service, but wants you to rent the telephone system from them as well. They may also offer Internet access. Their advantage is truly one stop shopping, but the overall cost to use a Shared Tenant provider is typically higher.

■ Final Thoughts

Find out which of these types of companies deliver service to your building and compare their service offerings, track records and capabilities carefully. Check references and don't lock yourself any long term contracts. Rates and service offerings change regularly.

Try to establish a rapport with your telecommunications service providers. The more they understand about your business, the better job they can do for you.

A Little Advanced Technology

*"Optional —
Only if you really want to know."*

A Practical Guide to DSL

The high-speed Internet access of the moment

What is DSL?

DSL stands for Digital Subscriber Line, a somewhat non-descript name for an exciting technology. The "D" is historical since the original form of DSL was a digital service. (Digital means that whatever is traveling over the line does so as either a "1" with current present or a "0" with current absent.) However, DSL has developed into a high-speed analog signal (usually represented by a sine wave) and is no longer digital. The "S" or subscriber refers to you or your company. You "subscribe to" or rent the DSL line from a telecommunications service provider. "L" or line means that this is an outside line (also called a circuit) that comes into your premise on a telephone cable that then runs out into the street and back to a telecommunications service provider. This is the same type of telephone cable used for your everyday telephone service.

The high-speed analog transmissions for DSL have different signaling patterns depending upon the type of DSL circuit and the type of hardware at either end. Two of the more commonly used DSL signaling patterns are CAP (carrier-less amplitude phase modulated) and DMT (discrete multi-tone).

The most common use of a DSL circuit is to physically and permanently connect you to the Internet so that you are "Always On." It can also enable you to connect to other locations (such as other offices of your company) through the Internet. With access to a DSL, you do not have to use a conventional modem with your computer and a regular telephone line to "dial in" (also called "dial up") every time you wish to access the Internet. However, there is a "DSL modem" (modem stands for – "modulate/de-modulate") that is needed.

One of the reasons that DSL is in demand is that it offers substantial capacity (also called speed or bandwidth), over a single pair of copper wires. Most home and office locations are already equipped with a spare pair of wires that come in on the cable that delivers the regular dial-tone telephone line. Therefore DSL does not require a new separate telephone cable. Since DSL is designed to use copper cable exclusively, it does so all the way from your premise back to the telecommunications service provider's central office to a device called a DSLAM or Digital Subscriber Line Access Multiplexor (Note: A DSLAM, pronounced "dee-slam", can also be located on-site in a multi-tenant or campus environment.)

The scenario described above using a separate pair of copper wires is known as using "dry copper." It is also possible to combine the delivery of your dial-tone telephone line for voice communications with a DSL on a single pair of copper wires. This is called delivering the DSL on "wet copper." You can have a telephone conversation and use your DSL at the same time, sharing the bandwidth (capacity) of the copper cable. Once this transmission gets to the DSLAM, the voice is separated and sent out over the public switched telephone network and the data on the DSL is sent to the Internet Service Provider' site.

There are some distance limitations with DSL, so your distance from the telecommunications service provider's central office will affect whether or not you can get DSL and with what speed. In general, the further the distance, the lower the speed you'll get, although speed is also a function of the hardware. If the DSLAM is on-site (such as it may be in an office building) then the distance limitation is not an issue. When the DLSAM is on-site, it is connected to another circuit to continue the transmission back to the telecommunications service provider, but the DSL circuit is only from the DSLAM into your home or office.

Types of DSL

▶ ADSL or Asymmetrical Digital Subscriber Line is most typically used for access to the Internet at home. The capacity of the circuit is greater coming from the Internet into the home (called "downstream") than going out in the other direction (called "upstream"). This is because home users are more likely to be receiving more information than they are sending. They may be receiving graphics, sound and video but the only things they "send out" may be keystrokes and mouse clicks.

▶ SDSL or Symmetrical Digital Subscriber Line is the type used by most businesses who need to send as well as receive significant amounts of information. In this case, unlike ADSL, the capacity of the circuit is the same in both directions.

The capacity, bandwidth or speed of the circuit (all terms used interchangeably) affects how quickly the screen on your computer will display the information from the Internet or how quickly you can send a "computer file" from your office to another location. If you are sending large files with pictures, you will need a higher capacity DSL circuit than if you are just sending regular text files or just using the circuit for e-mail. The capacity of the circuit also needs to be greater if several people will be using it at the same time (you can connect to the DSL through a computer server that is shared with other people in your office.)

Determining the capacity of the DSL circuit that's right for you may require a bit of trial and error. You can pay the highest rate and order the greatest capacity or order a lower capacity and see if it is sufficient. The capacity of the DSL can be changed by your service provider, usually within a few days.

Typical capacities are 128 Kbps (kilobits per second), 256 Kbps, 512 Kbps, 768 Kbps (a popular mid-range speed), 1 Mbps (megabit per second) or 1.5 Mbps. 1.5 Mbps is the approximate capacity of another type of telecommunications circuit known as a T1 (pronounced tee-one) that traditionally costs more than a DSL. (Most enlightened customers balk at having to pay for "tradition.")

The capacity, speed, bandwidth (pick your favorite term – if you really want to confuse your audience, bandwidth usually does the trick.) – refers to the volume of information that can travel over the circuit within a given time frame. 128 Kpbs means that 128,000 bits can pass a point on the circuit in one second. While the term "bandwidth" makes it sound like the bits are all passing through at the same time, they are actually passing one at a time in sequence – but it can be so fast that the end result looks like they are all arriving at the same time.

You may hear about other types of DSL which are all variations on the theme. HDSL (high bit-rate DSL) has a speed of 1.544 Mbps and uses either two or three pairs of copper wires, instead of one pair. IDSL (ISDN DSL) has the same speed, 128 Kbps, or a slightly higher speed, 144 Kbps, as another service called ISDN (Integrated Services Digital Network) which requires dialing into the Internet on a special type of line delivered on a pair of copper wires. DSL Lite is a lower speed version of ADSL. DSL Lite is also called G-Lite . RADSL is Rate Adaptive DSL that adjusts the speed of the transmission based on signal quality. VDSL is a "Very High Speed Digital Subscriber Line" (12.9 to 52.8 Mbps downstream and 1.5 to 2.3 Mbps upstream.)

Why Use DSL?

Most organizations and individuals who have been using the Internet for several years started out by dialing into the Internet service provider using a modem. The highest speed you can expect from a dial-tone line is 56 Kbps, but it is typically lower (slower) due to line conditions, despite the speed capability of your computer's modem.

Save Time

One reason to change to DSL is that you will not have to wait for a picture or large file being sent from the Internet. As web sites increasingly become equipped with audio and video, having the higher capacity to receive these clearly and in a timely manner is becoming even more important.

Save Money

Many offices still rent a separate outside dial-tone line from the local telephone company for about $30. per month apiece for each person with a computer to use for dialing into the Internet. These lines are paid for whether or not they are used and, in addition, may incur a cost per minute for usage. A DSL line for $150. per month can replace the 10 separate outside lines and be shared by ten or more people. Unless everyone tries to use the circuit at the same time or large files are being transferred, it is likely that all users will benefit from increased speed in using the Internet and sending e-mail and attached files. The range of pricing for DSL may be from $40. a month for residential use up to several hundred dollars a month or more depending upon speed and service level guarantees.

At home if you use the "wet copper" approach mentioned above and have a telephone line and DSL both delivered by your DSL provider, this will lower your overall cost as well because you do not have to have a separate telephone line for voice and another line for dial up data communications and Internet access.

Be More Responsive

Many of your customers have instant access to e-mail and may expect the same of you. With a DSL circuit you are always connected to the Internet and receive e-mail as soon as it is sent, enabling you to respond quickly to the sender. This may also enable you to take advantage of opportunities for new business before the moment passes.

In addition, you may look at doing your own web-hosting or providing more information to your customers with the "always on" presence you have with DSL.

Experiment with New Technologies

Having a DSL circuit provides the capacity needed to experiment with technologies such as the "web cam", enabling you to have a video conference while sharing documents with others similarly connected to the Internet (and similarly equipped with "web cams.") Applications such as this can improve communications with customers or among employees of the same company in different locations.

Even without the "web cam" for video, you may experiment with "Voice Over DSL" similar to Voice Over Internet Protocol (VoIP) which is a current "hot topic." You need the right telephone equipment for this which can convert your voice into the form needed to send it using "Internet protocol."

How to Get DSL

Who to Call

DSL can be ordered from many telecommunications service providers including your local telephone company (the traditional well-known company or one of the new competitors called CLECs or competitive local exchange carriers), your long distance company (who probably offers local telephone services as well) or an Internet Service Provider. Many are offering free or discounted installation to promote the service and this may include the DSL modem.

Also check with your landlord to see if there is an "in building" DSL provider. This is becoming popular in office buildings and residential apartments and condominiums. A lot of college campuses are now providing DSL service to the student.

Provisioning the DSL Circuit

In most cases, the pair of copper wires and associated cable both in your building and back to the DSL service provider is still rented from your traditional local telephone company by the company from whom you rent the DSL. There are exceptions to this as other telephone companies run their own cables into office buildings and build separate networks under the street.

As we mentioned earlier, some buildings are equipped for DSL service with a DSLAM housed in a secure location. This DSLAM is connected to your location by a pair of copper wires running to your office. Another high capacity circuit connects the DSLAM in your building back to the service provider's location, sometimes called their POP (point of presence).

Connecting to the Circuit

When the DSL circuit gets to your premises, you may need the help of a person who understands the computer (or computer network) to work in conjunction with the circuit provider.

If the DSL comes to a large office, it will be dropped off at the "demarc" (point of demarcation) usually in the telephone equipment room or computer room and must then be connected to a cable that runs to the desktop or more likely to a computer server, router or hub if the DSL is to be shared and used to provide Internet access from your office LAN (local area network).

If the DSL is delivered to a small office or residence and uses "wet copper" a device called a "splitter" will separate the telephone line from the DSL at the desktop.

In either case, a DSL modem is needed to convert the DSL signal into a transmission to be received by the computer. And an Ethernet interface is also needed. In the case of the DSL going to an individual computer, this is interface is called a NIC (network interface card). If the DSL is to be shared the Ethernet interface will be either a hub or a router.

If a server is used, you'll address security with a firewall which is hardware and/or software designed to prevent unauthorized access to your office computer network from the Internet.

For the home or small office, there is technology that provides the ability to have DSL at an increased distance from the central office.

There can be a considerable amount of waiting time (6-8 weeks) between the time you order a DSL circuit and when it finally gets up and running, so give yourself plenty of lead-time. And just in case the DSL goes down, keep the capability to access the Internet and your e-mail using a dial-tone line and your computer's modem as a back up.

Introduction to
Voice Over Internet Protocol (VoIP)

*Some view this as the future
of voice communications. Stay tuned...*

The Renaissance Of The Telecom Professional

"Listen up telecom professionals!" If you think the arrival of VoIP is the beginning of the end and that voice communications will become just another function of the IT department, think again. You may soon be needed more than ever. Read on.

What Is VoIP?

VoIP is short for Voice Over Internet Protocol. The first question that comes to mind is "What is Internet Protocol?" and the next question is "Why should a telecom professional care about VoIP"? The answer to the first question is that Internet Protocol is a means by which computers communicate with each other, so why shouldn't voice or telephones communicate using the same protocol? The answer to the second question is simply: applications.

Digital Telephones Currently In Use

Most of today's telephones working with a business telephone system take the human voice and convert it into a digital form (a combination of 1's and 0's) right in the telephone at the desktop. Your voice travels over one specific cable through the walls of your office space in this digital form. It ends up going back to the system known as the PBX, located in the telephone equipment room. This digital transmission of your voice goes directly to a specific location in the PBX called a port which is assigned to your telephone only and does not change from one call to the next. A port is a physical location on a printed circuit board, residing on a specific shelf within the PBX and housed in the PBX cabinet. Your voice is then connected through that port to another specific port associated with an outside line. The PBX, also called the "switch" connects your telephone to an available outside line and proceeds to complete the call when you are dialing out. (Every time you dial "9" you may be connected through the switch to a different outside line. Most PBX's have many outside lines coming into them or a T1, which is the equivalent of 24 outside lines). When someone calls you, the process is reversed and the switch connects the caller on the outside line to your specific telephone through the port associated with your telephone's extension number. The call is thus delivered through this process called circuit switching, closing a specific circuit to complete the connection.

IP Telephone Systems

The newer type of telephone system uses IP or Internet protocol to complete the call path. The transmission of your voice looks no different to the network than other information (data) being passed along on your corporate infrastructure. The term used is to packetize your voice which conjures up a nice neat little package, and indeed it is. In every IP telephone there is a device called a NIC or Network Interface Card (a small printed circuit board.) Your packetized voice is sent through the NIC onto the network using the same pair of wires and through the same routers and LAN switches as your data communications.

Unlike the telephones working with a PBX where a specific cable and port on the PBX are assigned to each telephone, the IP telephone can be moved and plugged into any data outlet on the network (either within a premise or at another premise) and its extension and capabilities will be recognized. Since IP telephones behave no differently than a computer on your network, all the established IT rules apply when managing IP phones on your corporate network.

Packet Switching vs. Circuit Switching

The IP call path is a logical connection through a LAN switch from IP telephone to IP telephone. Unlike a PBX, the call is not passing through a port that is permanently dedicated to that telephone. In other words, the packetized voice path does not pass through a call processor like a traditional telephone call passes through a PBX, but is delivered point to point by the network. The IP call processor serves the function of setting up and tearing down the call only. If you are on an IP call and the call processor fails, your connection will not be affected. Different manufacturers have different names for this IP call processor. Cisco refers to it as Call Manager. You may also hear the terms "Soft Switch" and "IP PBX."

The ability to use an existing data network connecting multiple locations for voice transmission using VoIP depends upon how well the network manages the existing data transmission. Voice communications have different requirements for getting through the network than do data communications and must be given priority since voice is latency sensitive. If you send an e-mail and the network is congested, the computer sending the e-mail can back off and retransmit. No one will care if an e-mail arrives ten seconds later than it would have on the first try. Voice packets, on the other hand, must get through the network on the first try 99.9% of the time to maintain the appropriate QOS (quality of service). Otherwise the quality of the voice transmission will be unacceptable or in some cases unintelligible.

Make Way – Voice Coming Through!

Thus, to use VoIP on a data network the routers and LAN switches must be ability to recognize the voice packet and give it priority over other transmissions. Cisco, 3com and Nortel, for example, all make devices capable of this prioritization and are continuing to refine theses capabilities. However, a solid assessment of your corporate infrastructure is truly the key to a successful VoIP implementation. Voice reliability and quality is only as good as the network you put it on.

Crawl, Walk, Run To IP Telephones

While IP is a standards based protocol for the transport of a voice packet, each manufacturer approaches the "set up" and "tear down" of calls differently. Therefore, telephones from one manufacturer may not work with the system of another manufacturer. However, IP phones that have been built to open standards offer the best hope of ever having interchangeable phones.

Corporations have chosen one of three methods to incorporate IP telephone systems; as an adjunct to an existing PBX, a migration from an existing PBX or into a "green field" in which no PBX exists. IP telephone manufacturers have made it easy for companies to employ these methods.

For example, PBX manufacturers such as Avaya (formerly Lucent) are hopping on the VoIP bandwagon by selling IP telephones that work with their traditional PBX's. This is accomplished by having a PRI (primary rate interface- a T1 with additional signaling capability) go from the PBX to a device called a voice gateway which converts the calls carried through the PBX on that PRI to an IP form. This enables the IP phones (they have their own unique circuit boards in the PBX) to talk to the non-IP phones and to make and receive calls using the PBX. In effect they are embedding a LAN switch used for IP phones within their PBX, in anticipation of a gradual transition to IP telephony.

It is these same voice gateways that enable IP telephones anywhere to communicate with non-IP telephones. VoIP can be sent over an ISDN connection (a dialed temporary high speed connection), a DSL circuit (most typically used for high speed Internet access) or over a Frame Relay network (used by many multi-site organizations for data transmission). Thus with an IP telephone, you can plug in anywhere and have the telephone appear to callers and to you that it is right in your office as long as you have established the connection to your office IP telephone system.

As mentioned earlier, when moving around your office you can take your telephone with you and plug it into any data outlet. Your extension and all capabilities of your telephone will be right there with you.

How Will We Use It?

Now that VoIP is becoming a reality, the applications making it useful are just beginning to develop. The early push for VoIP was to large organizations with multi-site data networks already in place. The pitch was that with VoIP they could use the data network for voice calls among locations and save a lot of money. With the plummeting of rates for long distance voice communications and the need for many corporate networks to be upgraded to take advantage of VoIP this has not happened as much as had been anticipated. Nevertheless, as new and improved ways of working are developed, we'll see increasing interest in VoIP among the average business users.

Development efforts enabling the merging of voice applications with Internet and data applications are numerous:

1. Someone calls your telephone number and the call rings simultaneously at your desk in the office, your desk at home and on your IP wireless telephone. The caller is sent to whichever telephone you answer.

2. At the airport wouldn't it be helpful to view arrival and departure information from the telephone sitting next to you at the executive club? If you then want to speak to an airline agent, your call will be prioritized since the system will know you are a preferred customer calling from their club.

3. You press a "service button" on your telephone and request information using HTML or connect to a database for on line

directory, calendars, local weather or attractions, company information to name a few. Twenty years ago, some telephone system manufacturers had phones with sizeable display screens, but they never took off since there was no Internet and therefore few applications. The big screen phones may be on their way back. Instead of the PC being used as a telephone (as many thought would be the case during the past 10 years) the telephone is morphing into a computer-like device that you can use to enhance the usefulness of your voice communications.

■ "Here You Go" Telecom Pros

One of the biggest factors in the successful widespread adoption of VoIP is the Human Factor. IT professionals are accustomed to taking certain parts of their computer network out of service or disabling devices to work on them. This can't happen in the VoIP world where telephones must always be working. So as the IT and Telecom departments merge, it must be acknowledged that traditional management of IT networks needs to change to accommodate the higher expectations that users place on voice communications.

Also (this is where telecom professionals can really shine) there are many areas not familiar to IT departments such as the handling of Interactive Voice Response systems, Automatic Call Distribution Systems (ACDs), dialing plans and ordering and provisioning services from the carriers such as PRIs, T1s, etc. Terms like ESF (extended super frame), B8ZS and clocking are also not part of the average IT professional's vocabulary. So there's a great opportunity here for learning and maybe finally getting the telecom and the computer departments on the same page. It appears that there's a lot to be gained!